Poison

Poison

A History and a Family Memoir

Gail Bell

St. Martin's Press ✖ New York

www.stmartins.com

ISBN 0-312-30679-2

First published in Australia under the title *The Poison Principle* by Picador, a division of Pan Macmillan Australia

First U.S. Edition: October 2002

10 9 8 7 6 5 4 3 2 1

For my parents, Roy and Alice – my gift
& for Andrew, with my love

Contents

I WILL NOT GIVE A FATAL DRAUGHT TO
ANYONE IF I AM ASKED,
NOR WILL I SUGGEST ANY SUCH THING.
from *The Hippocratic Oath*

IN EGYPT, A COMMON PRACTICE IS TO SCOOP OUT
THE CENTRAL PITH OF A CORN-COB AND FILL UP
THE REMAINING SPACE WITH ARSENIC.
John D. Gimlette, *Malay Poisons and Charm Cures,* 1923

I HAD READ ABOUT THE GIANT WATER BUG,
BUT NEVER SEEN ONE.
Annie Dillard, *Pilgrim at Tinker Creek,* 1974

Chapter One

The pleasures of found glass

WHEN I WAS A GIRL of ten my father showed me a
kind of sample case, made, he said, of the best wood,
lacquered, embossed with gold initials, hinged and
fastened with brass. When you undid the clip and lifted
the lid an inner shelf levered into view. On the shelf were
beautiful bottles each with a glued-on label, handwritten
in ink script, and sealed with glass stoppers. One bottle
had a thick sludge at the bottom, like tar. Idly, I picked it
up, extracted the top (which was stiff and unyielding)
and put the end of the stopper to my nose, as you would
with perfume. My father snatched it from my hand and
said '*Never, never* do that. You could die.'

The sample case had arrived in a carton delivered to
our house a week after my father's stepfather passed
away. The strange thing was, nothing inside the package
had belonged to the stepfather; it was what was left of
the belongings of my father's real father, and for some
reason had been hoarded, probably forgotten, until the
relatives tidied up and wanted to be rid of them. We laid

the other things out on a table: two leather suitcases from the 1920s, a collar box, a tie box, calling cards, a smoking jacket, a silver-handled walking cane and a small bundle of photographs. Looking at them seemed to give my father no pleasure at all.

'What's in the bottles?' I asked.

'Rubbish,' said Dad. 'Bad stuff, poison.'

Hearing this, my mother ordered the bottles out of the house.

I asked if I could keep the sample case but my father had made his decision. The next day he took it to work and buried it in the footings of a house. He put the other things back in the carton and put the carton in the shed where it sat for another twenty-five years, undisturbed.

>─┼─◆─○─◆┼─◄

My father worked on building sites for years and one of his pleasures was to uncover old bottles and bring them home. Clear ones with glass marbles in the neck, slender green ones that looked like schoolgirls, early Coca-Cola models that might be worth something one day. He lined them up on a shelf in his workshop where the sun caught their colours and threw them onto the walls. He showed me how to hold glass so its weight sagged in the cradle of my hand and warmed to blood temperature. Sometimes the delicate ones felt like little live things.

Often we cleaned them together, scrubbing at the clay with old toothbrushes and pliable pointed sticks my father had whittled to suit the work.

Years later I started bringing home my own glass: old tincture bottles, mortars, pestles, ampoules of sealed liquid, measuring cylinders, beakers, hollow tubing that could be melted and moulded on a Bunsen flame; all saved from the rubbish bins of pharmacies that were rushing into the 1970s without a backward glance.

Not once, in all those years of bottle collecting, have my father and I spoken about *those* bottles in the wooden sample case. He would say, if I asked him, that he couldn't remember which house he buried it under, wouldn't know if it's in concrete or fill, couldn't hazard a guess as to the possibility of excavation.

We've gone on appreciating the pleasures of found glass in this denying but comfortable togetherness that, for him at least, confirms his opinion that some things are best left buried.

When he retired from work he gave me the best of his bottle collection and sold or gave the rest away. He was done with dust collectors, he said.

I still line up the ones I like best — the green ones — on sunny ledges.

Chapter Two

Ingredients for a hell-broth

DECADES LATER, SITTING at my desk, I try to trick my memory into retrieving the names on those labels. My mind wants to see *Tinct. Nux Vomica* in light upward and heavy downward ink strokes centred on gummed paper with a gold leaf border. I want a small brown bottle of unusual shape labelled *Strychnine — Poison*, with a skull-and-crossbones warning symbol, and a crust of powder at the rim. The recall exercise, however, doesn't work. Recall demands some debt from the moment, some investment in future remembering that my younger self neglected to fund. To think, I say to myself, the evidence was once in front of me, preserved and unassailable.

If I were to come on the hoard now, if the house that sits on the sample box were torn down and a shovel should catch its brass handle with a treasure-finder's *ding*, I would pounce on those bottles like an Egyptologist on a funerary mask. In a flash I would know what sort of conjuring trade my grandfather practised, how far his experiments with formulation had taken him, how far from formal training he'd strayed.

To my father the names on the bottles were inconsequential details; he wanted no part of this confounding, unsolicited legacy. A piece of the past had lurched out at him like some sort of cellar beast that had to be fought back into its vault. Historical interest didn't come into it, neither did curiosity or sentiment. I was ten, he was thirty-three. We stood in exact relation to the ages of himself and his mother when she died, and *her* death, not his father's, marked the boundary of his stake in the past.

▷─◁▷─○─◁▷─◁

By inclination and training I've spent many years thinking about the bad stuff, poison, and its companions: secrecy, death and storytelling. The obvious things about poison — what its many forms look like, what happens when it mixes with body chemistry, who used it, who died, who lived — are part of the meditation because it couldn't exist without them; but the longer I think my way into poison's secrets the more I retrieve the buried shapes of other things — ghostly suggestions of poison as an annihilator, a molecule rearranger, a test of defences, a harsh father, a wise mother.

As a death-bringer, poison can be merciful (if it arrives in the company of surprise, swiftly and cleanly on the point of an arrow), but for the most part, poison works slowly, under the cover of the body's regular failures of health, and there is no charity in its hidden agency.

Thinking and reading about toxicity is — at one level — an outgrowth of the work I do; at another more compromising level it's a way of meditating on the man who owned the wooden sample case, my grandfather, William Macbeth, chemist and poisoner.

The story of my grandfather comes from the out-of-bounds territory of our family lore: the dead zone of secrets, severity and things withheld. It was passed to me when I was a young woman finished with study and curious about connections to the past, and, in the way we did things, I was meant to read the warning label and put the story away somewhere out of reach. William Macbeth was the grandfather who died before I was born, the one we never spoke about. To have him unexpectedly break into the present moment of a life and shape a space where there had been none before brought me to what became, in time, a persistent intrusive interest in the way poison stories are told.

William was not so much a skeleton in the family cupboard as a walled-in ghost pacing his few square feet of cellar floor. His poison was strychnine, the bitter seed of a nut that grows in tall trees. He mixed strychnine tincture into cordials to mask its sharpness and gave it to two small boys, his sons, who died the awful, convulsing, diaphragm-contracting death of a paralytic poison. The story was talked into its final shape by William's sister-in-law, Rose, and left to lie, unexamined, in the occasional thoughts of future generations. The unexamined version became the definitive version because no other existed.

Two kinds of truths seem to emerge from the large canon of poison stories. The historical truth, the one that's constructed *after* the event using science (and, more recently, psychology) to arrive at a resolution that fits the evidence; and the narrative truth, a literary telling which is contiguous with the crime, precedes it, survives it, and lives on afterwards in a sort of background murmur. Sometimes, stories are a melange of subjective truths, and sometimes lies tell you more than all the evidence you can hunt down.

⊱──◦──◦──⊰

My desire to see the man in the monster, William Macbeth, was the beginning of a long journey which started with what I knew about poison (from science) and moved outwards to what I needed to know. Much of the work had already been done, some of it most successfully, it seemed to me, by playwrights and poets. Poison stories steal their power from mythical figures and symbols, the same figures who elevate the great stories and speak through them in intelligible ways about mystery and fate. The gaze my training sharpened in me, the one that worships the solid virtues of facts and is drawn from one to another and another looking for better and brighter data, acknowledges the apparently different territories that literature and science have laid claim to, but is reluctant to concede that one knows better than the other, merely that each reads the landscape with different instruments.

In anthropological science I found an image with dramatic possibilities. To read the image, you have to think your way into an early twentieth-century Malayan jungle, at sunset. The tribesmen are making blowpipe poisons. Into the pot they throw ingredients for a hell-broth: centipede heads, scorpion stings, fish spines, snake venom, red arsenic, sour lime juice, opium and a dash of saffron and spice. They tend the brew according to custom, exactly so much of this, barely a dash of that; they stir until it thickens, clockwise so many turns, counterclockwise so many turns. The ritual is practised, well rehearsed, and hierarchical. When the tribal chemists step away from the pot, the magicians step forward.

The magicians have been warming up with handclapping and chants. As they speak their rhyming spells and move their carved sticks through choreographed arcs, all eyes watch for the moment of transmutation, when the essence, the indefinable something that is both unknown and recognisable, reveals itself in the bubbling mixture.

The magicians know what is going on. They have a Malay name for it.

Shakespeare has Hecate, queen of witches, call it 'Enchanting all that you put in.'

The Western observer who first recorded the Malay ritual called this moment, laughing behind his hand, 'the projection of the poison principle'.

Much of what follows is a journey around that pot of simmering juices.

Chapter Three

Bitter seeds

AH, MY POOR PRINCES! AH, MY TENDER BABES!
MY UNBLOWN FLOWERS, NEW-APPEARING SWEETS!
IF YET YOUR GENTLE SOULS FLY IN THE AIR,
AND NOT BE FIX'D IN DOOM PERPETUAL,
HOVER ABOUT ME WITH YOUR AIRY WINGS
AND HEAR YOUR MOTHER'S LAMENTATION.

Richard the Third, Act IV, Scene iv, 8–13

IN 1980 ROSE CLARK, my great-aunt, permitted a record of our family's poison story to be written down. She was an old lady who would die within six months of the telling and though this was our only meeting, and she was reluctant, I think she believed her motives were true. I liked her wiry, hard-knocks figure, her cardigan in summer and the way her hearing sharpened or dulled depending on the colour of my question. I judged her to be honest even if her motive — articulated as 'speaking her truth' — appeared to lean in the direction of score-settling.

Most of my family knew the outline of the story by then: my grandfather, William Macbeth, herbalist and chemist, poisoned two of his four sons with strychnine. Because the family lived in fear that he might strike again if challenged, and taking for granted that the community would not believe ill of one of its trusted identities, he was never reported or charged. After a time, to protect the surviving children (according to Rose), his wife Ellen left him. She moved in with her older sister, then after a short illness Ellen died leaving Rose the burden of extra young children, who were eventually placed with another family remote from their father. Some years after the murders William success-fully impersonated a doctor at a lunatic asylum, briefly achieving recognition and wealth by plundering the bank account of a rich inmate, before dying of tuberculosis, alone and unreconciled with his children (one of whom is my father), in 1948.

I counted myself lucky to have an eye-witness willing to talk. I sat silently and respectfully with my notebook and pen taking everything down, only speaking when I wanted clarification of a word or place name. During a tea break I wrote, 'Room dark, curtains drawn, can't see her face, back-lit, in shadow.'

Her daughter brought the tea. Clearly she disapproved of raking over the past. Rose's other sisters, Edie, Ivy, Jean and Betty, had all declined to be interviewed. Ivy wrote, 'I can't see what you think to gain by asking'; Edie and Jean pleaded to be left alone with their memories of poor dead Ellen; and Betty pleaded ignorance, being the child of her parents' middle years and too young to know anything of the events of 1927 first-hand.

So this was a gratefully accepted chance to speak to

someone who had breathed the same air as the family monster and survived.

Rose didn't miss a thing and was splendidly generous with her rendering of detail. She began at the first meeting between her impressionable younger sister Ellen and the tall, handsome William, which occurred in Orange, a country town west of Sydney. No chance meeting, the sisters were in fact waiting for 'Doctor Macbeth' to arrive, along with fifty or so locals, in a hall set up for a night of 'medical miracles, faith healing and marvellous cures'. Orange was the final stop on an extended tour of rural New South Wales for Macbeth's travelling one-man show.

When the healer stepped up to the lectern wearing a silk top hat and long cloak , several ladies 'blushed'. Rose and Ellen were among the first group to take part in a healing circle. Rose suffered with asthma, Ellen had come along out of curiosity. Macbeth moved around the circle touching their bowed heads. With some people he pronounced immediately, with others he was silent as if waiting for guidance — and in one case he took a young man aside to speak to him privately. When his hand brushed Rose's hair she experienced a sense of suffocation followed by blessed relief in her airways. When he faced Ellen, he firmly cupped her chin in his hand and smiled.

Afterwards he sold cures and presented business cards which described him as *late of Pennsylvania, U.S.A., Doctor of Naturopathy, Herbalist and Botanist, Dispenser and Medical Masseur.* His showpiece cure was *Macbeth's Strengthening Tonic (tested in the highlands of Scotland on descendants of the great warriors of Culloden).*

William and Ellen married soon after that meeting, in June 1922. Rose submitted no linking sentence between the meeting and the marriage. Photographs of William show a tall, thin, angular gentleman with sandy hair, short sight and expensively tailored suits. In one photograph he is posed for clay pigeon shooting in jodhpurs and high boots. Ellen's image shows a sweet-faced girl (she was eighteen at marriage) in a home-made dress. Their first child, Thomas, was born with physical and mental defects.

William found them a house to rent in Katoomba in the Blue Mountains west of Sydney. Ellen's sisters, Rose and Edie, moved in temporarily to help. William's travelling show went off the road and three more sons were born in as many years.

By invitation, when they were in town, Rose and Edie assisted William with corking and labelling his tonics, a job that made them feel 'important' and 'needed'. Rose, the eldest, took to the role of apprentice with a certain amount of high-mindedness. I glimpsed the sorcerer at work through her sharp eyes. She noticed everything William did, and thought there was 'a lot more to it than bottling quinces'.

William's habit was to work from left to right. On his bench he kept a mortar and pestle, a lot of 'tools', and a travelling case with compartments for bottles and jars. He mixed batches of tonic in a vat that had to be washed with a weak spirit solution and rinsed in boiled water. He collected empty bottles which Rose cleaned and lined up 'like soldiers waiting their orders'. Edie concentrated on Ellen and the children.

So what happened to alter this picture of the extended family rallying in a time of crisis? How do we get to that point in the narrative where the mood changes and the 'but then' clauses appear, the change in breathing that signals a new direction?

Rose extracted two events from the flow of recollection and laid them out like cards. The first was the breakdown of the primary relationship between William and Ellen (accelerated by a sick child, too little money, and William's enforced staying-put). The second was the arrival of the poison, strychnine, needed for the latest recipe, *Macbeth's Tonic for Gentlemen Only*.

Rose's hiatus phrase was 'one day'.¹ One day William dosed his eldest son with the strychnine-laced tonic and the boy died.

Thomas, the child in question, has been described in various ways as a hopelessly 'retarded' boy of four with no speech and 'spastic' limbs. His parents knew he wouldn't live long, they'd been told to expect him to go one day, probably in a fit or by choking, so when he stopped breathing 'conveniently' at a time when the family was desperate for relief (and Rose and Edie were absent) no alarm bells were triggered. Ellen, said Rose, accepted the loss.

But not many months later, the three-year-old boy died too. Rose was back in town 'for that little homicide'. Patrick was a wilful child who trespassed in the dispensary. One day (again) the boy drank tonic, at his father's instruction, to be taught a lesson. He ran off to play by the fire, took a fit, was carried next door by his frightened mother, was given an emetic by a neighbour and died in the neighbour's arms. The detail about the emetic is new. Rose remembered that a neighbour had given the boy salty water, a well-intentioned but mistaken way to fight the excitatory effects of strychnine.

The Macbeths left Katoomba, worked at their marriage, separated, then divorced.

Later, with Ellen dead, William out of the picture and the two younger children in care, Rose and her sisters became the owners of the story.

><-<>-•-O-<•><-<

Ellen's sisters (except Edie) lived to be old women. They were deeply private women, but my family's poison story comes from their lips. They created a shape for these events, stitching their patches over the holes, strengthening the seams, snipping off threads, and even putting their needles to a bit of embroidery.

The murder version, Rose's version, seems to me to be lacking in motive. What did William have to gain by killing two sons? Why was Rose so sure Thomas died of strychnine overdose and not during an organic convulsion — did she know what a death by strychnine looked like? In my notes there's a sentence that says she'd seen a farm cat take a strychnine bait, a sight she wasn't likely to forget.

Rose's conflated memories, fifty years old, seemed to wander around looking for a direction before gathering their wits again at the edge of the cauldron. She hypothesised William the sorcerer testing his toxic brew on his sons: sacrificing a boy destined for the grave anyway, then compounding his sins by killing the three-year-old for sport, or even spite.

I asked Rose to consider another version of events, the 'terrible accident' version. What if Patrick got into the dispensary, found the sweet cordial, tasted it, gave some to his sick brother and the sick one died because of a weakened defence system? Time passes, he wants more, he overdoses himself and passes out. The wrong antidote is tried, he dies too. What about that for a story? It absolves the father of deliberate murder, it explains Ellen's passivity, it cleans up the family history.

In fairness to Rose, the calamity version wasn't news to her. More than that, she'd nagged her conscience over it, but in her mind the deaths of those two boys all came down to the same thing — if William hadn't been experimenting with poison her nephews would be alive today. And on this digressionary path Rose laid down another version, the 'expedient plot' with its fifteenth-century echoes of two princes in the tower.

The eldest boys, one sick, one wilful, stood in the way of William's ambition. The patent medicine business made money, but never enough. William with his quick mind saw a chance to open out his market with a strychnine sex tonic, something he'd read about but had no dispensing experience with. In an act of pure treachery, he dosed the sick boy to see what would happen and what happened was so unexpectedly beneficial for all concerned — for the boy who was now out of his misery, for the mother who had reached the end of her tether, for his own chances of really achieving some progress with his work — that he kept quiet and let them think nature had come to the rescue. He gave nature another helping hand with Patrick. The coroner exonerated him because he was educated and could plead his case 'so charmingly'.

Retreating only slightly from this extreme version, Rose summed up by saying that William was a murderer, whichever way you wanted to twist the words.

When poison stories get going (written, reported, discussed), language changes shape. Descriptions undergo a tidal pull. It seems difficult to hold your course once you get close to the main characters and the poison vessel. The skew is towards hyperbole, melodrama, and extremes. Rose's story from 1927 (retold in 1980) opposes the rakish, rather dreamy William against the passive, too-fecund Ellen. There are innocent babies and a crippled boy. Rose calls William a 'scoundrel'. One minute she's thrilled to be asked to stir the cauldron, the next she's suspicious of the brew. William is both a healer and a killer. Even my own description of my grandmother and her sisters as 'sweet-faced' is manipulative.

If this story had been painted it would be rich in iconography: William in his dispensary (a converted cloakroom) with its dark compression of space, the shadows, the single electric bulb giving a lemony wash to the light, his tools and bottles, the labels and corks arranged each side of the mixing vat. Our eyes would be drawn to the vat, the crucible at the centre of the picture, and to the hand which adds the poisoned tincture, drop by drop. We might imagine the poisoner's journey: turning the knob, balancing a tray of washed bottles, needing a foot to guide the door shut, stepping to the workbench, taking up the rhythms of stirring, blending, adjusting to volume. Secrecy, planning and a certain amount of manual dexterity are needed to hide poison in a bottle. What if it doesn't dissolve? What if you can see lumps, or grit?

Gazing on this picture we are conscious of an implied person or persons for whom the poison (or spell) is intended, in William's case the two young boys in another room. The victims, innocent or otherwise, are so strongly bound up with the mixing vat it can have no other purpose than to be cooking up something to give to someone. Getting further into the imagination of this picture we might be present at the actual moment when William, in the form of his whole self — father, chemist, husband — splits off into the murdering self. What would it look like? something as obvious as a glint of teeth? A Dr Jekyll/Mr Hyde horror show?

Painters have tried to capture this look. In my study I have a print of J.W. Waterhouse's *Circe Invidiosa*. It gives a glamorous rendering of the sorceress Circe at her poisoning work, dressed in a peacock gown: her gaze is subtle, intelligent; whatever is going on is subterranean and no business of ours.[2] And yet we look.

William's other-self, the one who kills the boys, is not a subtle or grieving self. There is an earthy, underworld smell like buried iron on his fingertips and a cauldron stink to his cloak.

Our eyes linger on the mixing vat because, whether we know it or not, it's an alchemical icon, a symbol of what we unconsciously know about transformation, rebirth, and the devouring feminine archetype. The vessel as metaphor is the uterus in which the seed is planted and from which alchemical gold is harvested. Not so clearly described are the ingredients inside. Riddles, puns, metaphors, tricks of language and nomenclature hide the identity. No one has yet named the *prima materia,* described it scientifically or held up a specimen. Alchemists developed a secret, hinting language to disguise its face. Quacks like my grandfather borrowed freely from alchemy. The wizard archetype is a useful brand identification trick; and many quacks, William included, had a flair for Paracelsian self-promotion and a belief that medical wisdom (*sapienta*) came from selling your message on the road amongst the people, and not from books.

William borrowed the rituals, the secrets, the disciplines (his habit of working from left to right) and adapted them to the dynamics of getting into profit as quickly as possible. There was no time to master years of study, solve riddles, interpret allegorical writings, no desire to fast, or undertake strange diets, or build three kinds of furnace which might blow up in the second distillation phase.

From family gossip I know that William's *prima materia* was Beecham's Pills, a commercial compound of ginger, aloes, coriander and the oils of rosemary, juniper, licorice and capsicum. He ground the pills to a fine powdery state, dissolved the mass, filtered off the residue, added syrup, green food colouring and whatever extras he had to hand. Real alchemists tried substrates like horse manure, boiled eggs, sea salt, dew, alum, sulphur, mercury, silver, iron, antimony and the one we all know, lead.

Lifetimes of labour have gone into torturing the essence from base ingredients in the hope of striking the seed of transformation. Without it, no other step can be taken. Whatever goes into the vessel is always putrefied first to the black *nigredo* form: this is the stage of particle rearrangement, breakdown and chaos. Devils are let loose. In Jungian terms, the controlling conscious self surrenders to the uncontrolled unconscious self — perforations in the boundaries allow leakage from one side to the other. Anything might happen. I think this is the prefigurative moment every poisoner reaches. It might be as quick as a blink.

Fairy tales always use perfect poison. It is absolutely

not helpful to the narrative to have a clumsy chemical gushing straight through the victim's insides and out the other end. The Old Queen in Snow White 'made a poisoned comb, by arts which she understood', which felled Snow White with a single stroke. This is the perfect contact poison in action, one brush on the scalp and its potent force subdues mind, body and spirit, bending and collapsing the victim into a swoon that, in this case, mimics the final annihilation of death. The Queen certainly thought Snow White was dead, but hadn't reckoned on the deft attentions of the dwarfs, who removed the comb so that Snow White recovered without side-effects. Finally the Queen 'went into an inner secret chamber where no one could enter, and there made an apple of the most deep and subtle poison'. This symbolic descent to an underworld, the darkest corner knowable, is a journey made by every poisoner. It is a journey to the shadow character we are all said to possess, that place in our personalities where our own unacceptable characteristics collect. It is also a journey to a real place, like Macbeth's dispensary.

Snow White doesn't suspect a thing, she seems to revert to a childlike innocence without her dwarfs nearby to function like her senses of taste, touch, smell, sight, hearing and the all-important sixth one which ought to warn her against old ladies bearing apples.

⊱—⊹—○—⊹—⊰

Poison-story language borrows its arousing, emotive adjectives from fairy-tale polarities of good and evil. The poisoner character is wickedly larger than life. He's a trickster, a shape-changer, a cunning fox who keeps us amused as we walk the death path. Often, he dresses the part. From a newspaper story of 1932, by which time he was a *divorcé*, I learned that there had been a great deal of fuss about my grandfather's clothes.

At a time when Australia was in Depression and people (like Rose) were wearing dole boots, William owned 'twelve pairs of shoes, half a dozen day suits at 25 pounds each, two evening suits at 30 pounds each, six shirts, a presentation watch, six firearms, six hats, a car valued at 760 pounds, a gold cigarette case and gold card case. His matching leather luggage pieces were all gold embossed.' He also had £5000 in a bank account, a chauffeur and a 'pretty brunette' girlfriend who liked guns. His chauffeur testified: 'He told me on more than one occasion that he was a qualified medical man, and that he gained his degrees in America. Also that he was an American citizen educated at Oxford and Cambridge. Further, he told me that he had practised in the country as a doctor of bloodless surgery and extreme phthisis, and in one country town had practically put a famous doctor out of business. He had made such marvellous cures that he had accumulated enough money to retire on for some years.' Rose's most cutting remarks about her brother-in-law always included a swipe at his flashy clothes.

My expectation that Rose would guide me to a deeper understanding of the boys' deaths was optimistic, and even unreal. In 1980 I was hooked on the sensationalist aspects of the story. My curiosity was reactionary; if my father wouldn't probe the wound then I in my fearlessness would front the old lady and get some answers. Satisfied, I walked away from it. There was no second interview, no backtracking.

Yet the sense that something had been withheld or

misappropriated by Rose never really went away; if I attempted a mental reconstruction of William giving strychnine to his children from the insider perspective of my own work in a dispensary — even within the broader parameters of the maverick professionalism of a quack — the furniture didn't fit. Ellen never really comes alive in the story, she's been sidelined by Rose, yet Ellen is the hearthstone, the mother who lost two babies, the betrayed and grieving wife to whom bad things happen, and then she dies. Then there's the ugly but necessary piece that won't fit anywhere — the motive. Rose and I shifted it around between us but it still sits stubbornly blocking the passageway and has to be climbed over. Finally there are the magician's clothes that seem to lie about everywhere obscuring shapes and views. I look at pictures of my grandfather and see not even the rippling simulacrum of a monster.

Ten years after Rose died (her story with its score-settling and imperfections stitched firmly into the lining of the family jacket) I interviewed the surviving sister Betty. She, like me, had the received version of the story. She'd never met William, was twelve when Ellen died and viewed Rose (twenty-five years her senior) more like a parent than a sister. Betty's only original contribution is an afterthought, something she picked up from weary listening: 'Rose fancied William,' she told me, 'I think that was half her trouble.' This was all Betty gave me yet it was my first clue.

⊱┈✦┈◦┈✦┈⊰

Of dying by strychnine, I knew a little. Of the power of poison to subvert a narrative, I was learning as I went along. At this early stage of a long trawl through the written and spoken records of poison stories I turned to what was written of strychnine, working magpie-fashion through the published tracts, my eye alert to possibilities.

Chapter Four

The victim begins to twitch

A MISTAKE IN terminology originally put strychnine in the category of 'emetics'. The name of the seeds of the Koochla tree, *Strychnos nux vomica*, was transliterated as 'vomit nut' or 'emetic nut'. Actually, 'vomica' means cavity or depression. The cavity is slit-like and is the subject of storytellers: legend attributes the depression to the digital imprint of the Creator. But strychnine doesn't make you vomit, it convulses your spinal cord.

The English and Germans first used the bitter seeds as a rat poison in the sixteenth century. The seeds were very effective, though nobody knew exactly why. About 1540 it was written into the European pharmacopoeia, but it wasn't until 1817 that scientists isolated strychnine as the active drug in the seed, and two years later a second (but less toxic) alkaloid, brucine.

What you get out of the seeds after a bit of chemistry is a translucent crystal which can then be ground to a white gritty powder. There is no odour but a very bitter, metallic taste. You can dissolve the powder in 400,000

times its own weight of water and still taste the sharpness, hence its introduction as a stomachic, or tonic, for
ailing appetites. The bitterness was supposed to irritate
the taste buds and the stomach lining so they'd run with
juices, making you hungry.

Medicinally, the drug became available in a number of
forms, some of which can still be found on the back
shelves of modern dispensaries, and in homeopathy.

When taken in overdose strychnine is a paralytic
poison. It rapidly travels from the gut to the blood and
from the blood to the spinal cord and brain. The spinal
cord becomes excited and surrounding muscle groups
begin to contract. These contractions begin in an orderly
sequence and end in chaotic convulsions, with the
stronger muscle groups dominating. The victim begins to
twitch.

At the same time, the brain's cortex is attacked.
Studded over the cortex are the control centres for our
senses. Ravaged by strychnine, taste, touch, smell, sight,
hearing are dramatically enhanced as if all the brain's
power were diverted to serve their tiny radar stations.
Every signal, every nuance, every atmospheric distur-
bance in the environment is picked up, thrown onto a
screen and enlarged in wild computer-enhanced detail,
making a touch seem like a blow, the brush of an eyelash
a whipping with birch stems. Something small can send
the whole body into convulsions; the trunk arches
backwards until the only parts touching the ground are
the crown of the head and the heels, the jaw and fists
clamp shut, the eyes protrude, the mouth muscles are
pulled into a grimace (*risus sardonicus),* the diaphragm
contracts, breathing stops for one minute, maybe t vo,
then the convulsion ends. If still alive, the victim begins

to breathe again, is flaccid, exhausted, oxygen-starved and full of fear.

Death often follows the second convulsion which erupts about ten minutes later, but death may be delayed until the fifth convulsion.

In the objective space in which one comprehends medical description and struggles for clinical detachment, the events follow a neat logic. A is followed by B is followed by C. Few of us have ever watched a human death by strychnine, and if we could what we would see would be superficial; we cannot see the internal seizure, we cannot climb aboard a fibre-optic probe to follow the A B C that leads from a writhing spinal cord to the final halting of the medulla, at the base of the brain. We can't follow the beast and observe its habits, can't witness it breaching the body's defences, consuming its host, cutting off its oxygen supply.

Using empathy and imagination we are able to enter the parallel universe of subjective experience. We can sit by the fireplace with three-year-old Patrick, William's second son, and stare into the flames, our muscles tense, our eyes beginning to bulge, our gaze captured by the sight of dancing red air. 'The muscle contractions are quite painful and the patient remains acutely apprehensive and fearful of impending death throughout the ordeal,' says a standard pharmacological text.

Patrick stopped breathing during the second seizure when a well-meaning neighbour administered a salt and water emetic. The 'well-meaning' qualification is there because making someone spit up the bad stuff is standard practice for many poisons, but it is the very worst sort of help Patrick needed. Being made to swallow salt and water was tantamount to killing the boy more quickly.

The aim in reversing strychnine poisoning is to protect the patient from all stimuli, to lay him in a dark, quiet place and give him a muscle relaxant. A little inhaled chloroform might have saved him long enough to procure chloral hydrate or paraldehyde, the best antidotes available in the days before barbiturates or benzodiazepines.

In clinical doses, strychnine has a well established, though now nearly redundant, place as a stimulant — of circulation after surgical shock, of appetite (the bitterness), of homeostasis after depressant overdose and, anecdotally, of virility. The Victorians were great fans of the strychnine aphrodisiac. Dedicated users developed a tolerance to its toxic effects and suffered withdrawal-like deprivation if the chemist was uncooperative. In 1989 the American FDA put strychnine on its list of banned aphrodisiacs. 'There is no conclusive evidence demonstrating the effectiveness or safety of any plant materials that have been used historically for aphrodisiac purposes.'

There is not just one variety and one type of bitter strychnine seed. The Malayans know a high-climbing woody creeper, *Strychnos tieuté*, whose globular berries contain silky coated seeds that are boiled up for dart poison. Interestingly, these fruits are eaten by the local monkeys who are completely immune, just as in domestic veterinary practice the poison is safe for some pets. Strychnine dog tonic, a 'time-honoured' preparation from the 1950s, improved general body tone and promoted a sleek and shiny coat. Its ingredients were iron, ammonia, tincture of *nux vomica* (strychnine), arsenic and chloroform water. 'There is no doubt that many dogs are benefited metabolically (for reasons unknown) by an arsenic and strychnine tonic,' says a veterinary manual of the 1960s. The dispensing trick was to shake the bottle well before dosing, because a concentrated sediment formed at the bottom of the bottle and a dose of this sludge *could* poison a dog.

The five main characters of my family murder story are all dead. The Brothers Grimm have a story about the Godfather Death. There's no bargaining with him, no mocking his power. He strides on withered legs between his underground kingdom and the world above, checking on things. It's a wearying business. Downstairs he keeps a lamp burning for every living soul. In the normal span of things, children have tall dancing lamps and old people have small flickery lamps, and as lights go out (as they do every minute of the day) others flare up to take their places. Now and then a child is born with a low flickery flame and Death knows to have his boots ready for a trip to the top.

I wondered, did Godfather Death make two separate trips to collect William's sons in 1927 or did he save his shoe leather and hang around a bit, waiting.

The answer depends on who's telling the story.

I needed to listen carefully. Lies carry buried truths. The search had to broaden. I needed to see the face of the poisoner, study the context, listen to the language (mine too), note the auguries, and contrive to see the dark shape in the mixing vat, the poison principle itself.

Chapter Five

Post mortem courtesies

. . . HER PHYSICIAN TELLS ME
SHE HATH PURSUED CONCLUSIONS INFINITE
OF EASY WAYS TO DIE.

Antony and Cleopatra, Act V, Scene ii, 348–50

HERE IS A TALE OF POISON:

Once long ago the Queen of Egypt laid out plans to die by her own hand.

The Queen had a vain heart. She wanted no wounds in her flesh or grimace of pain on her face.

'Are there drugs,' she asked her physician, 'which bring on painless death?'

'Of course,' replied the learned man. 'I will prepare doses of my best poisons.'

He demonstrated his drugs on slaves who were brought to a special chamber, but nothing the Queen saw matched her idea of a perfect death. There was too much shaking and crying out.

She asked to see other, less common, potions. The physician begged leave to visit certain women of the city and later that day presented three vials for consideration; in one was a liquor of hemlock, in another sweet-smelling laurel water and in the third a quantity of exotic seeds. He laid out the virtues of each potion; how many drops, the best wine to sweeten the taste, the most auspicious lunar phase in which to drink, and for each, the correct incantation.

'Show me a corpse from each,' said the Queen.

Two prisoners were chosen. One was given laurel water, the other hemlock. The court waited and watched. When one prisoner twisted on himself in a slow death struggle and the other turned blue at the mouth, the Queen became impatient and asked to see the manner of death from the third vial. So the doctor pulverised the seeds, made a tincture with wine, saying (as he had been told) that with this drink death comes quickly to the call. The Queen showed great interest in the demonstration. Minutes passed with the prisoner standing still and straight.

The physician flew into a panic, prepared another draught — doubling the dose this time — but before he could put it to the man's lips a sudden terror took hold of the prisoner and he threw his body onto the floor in a series of terrible fits. He died with his mouth drawn up in an ugly smile.

The Queen lost patience. 'Speak,' she said, 'about arsenic, white lead, quicksilver.' The physician shook his head. 'Regretfully the minerals are slow killers and not the wisest choice when the errand is urgent,' he said.

So the Queen sent him to the snake handlers.

To get fresh young cobras the doctor went to the

riverbeds himself but he was old for this sort of work. When he returned the Queen had her maids assembled ready to listen to the news. Battling tiredness, the physician began an oration on the relative merits of cobras and vipers. 'A viper kills by poisoning the blood,' he said, 'and there is much that is *not* pleasing about a corpse that is viper-bitten. A cobra, on the other hand —' but the Queen cut him off in mid-sentence.

She ordered prisoners to be brought to test cobra venom, and being satisfied, instructed her attendants in certain *post mortem* courtesies. A Queen, especially in death, needs repaired make-up, a closed mouth, and combed hair.

When the hour of her humiliation drew near, the Queen begged time alone with her maids to prepare for the afterlife with prayers and incantations. When she judged the time was right, she sent out for a basket of figs which concealed a snake.

Dressed by her maids in ceremonial robes and a golden crown, the Queen lifted the serpent and caused it to bite, twice. She lay down while her maids tidied up.

Alerted by the guards, the enemy king rushed to the tomb to find his great prize, the Queen of the Nile, lying dead. Her face shone with gold dust and pigments, her expression was serene and majestic. One of her maids was also dead, the other on the point of dying.

'This is cunningly done,' said the enemy king, who had to say something to hide his fury at being cheated.

➤┤⟐⟐•O•⟐┤◄

This is one of many versions of the last hours of Cleopatra, a story that has been lifted out of history and into myth by frequent handling over the centuries.[3] What is different in this telling is the addition of forensic detail — the cyanide and strychnine experiments; the 'cobra not the asp' theory; the requirement that an attendant tidy up the body after extinction of life. Some accounts of Cleopatra's last days suggest she was looking for a drug (the sort Juliet found centuries later for *her* tomb scene) which would bring on a pseudo-death then wear off when the coast was clear. Others suggest that Octavius deliberately left Cleopatra alone hoping she would suicide.

Whether the Queen of Egypt achieved her wish of 'easy' dying is unknowable. When the body was examined all that was found were two small puncture marks in the skin. Theories about how she stage-managed this ultimate exit began on the day she died and continue to exercise the forensic interest of scholars. There are theories about hollow tubes containing arrow poison hidden in her hair; poisonous snakes swimming in the water jugs; maids trained in deadly martial arts.

Faced with the basket of figs and a glimpse of something moving, Octavius fell in with the asp theory. One story has him calling in the *psylli*, a group of Egyptian snake charmers famed for their immunity to serpent bites, summoned (too late) to suck out the poisoned blood.

━┼═◈══○══◈═┼═

Poison stories are a subset of murder stories. In British law, poisoning could never be tried as manslaughter because the offence was always considered to be deliberate and premeditated.

'Cunning is an essential element in the successful

poisoner,' writes an expert in toxicology, 'and the exclusion of any sense of pity from his make-up is an inestimable asset, since he has to witness the results of his handiwork and watch the life of his defenceless victim slowly drawing to a close.' Framed in the language of authority, the poisoner's crime is elevated out of the run-of-the-mill homicide category to a hyper-real place that smells slightly of witchcraft.

Despite the masculine pronoun, the poisoner's crime has always had a feminine profile. The commonly applied adjectives 'beastly', 'direful', 'secretive', 'furtive', 'foul', 'vile' 'clandestine' (which seem to come from pens dipped in mossy inkwells) characterise the crime as *weak* and *recessive* — as opposed to *strong* and *dominant* acts like shooting and stabbing. Professor Glaister, in his 1954 classic *The Power of Poison*, opened his first chapter this way: 'Murder by poisoning is a crime of devilish wickedness and inhumanity which no language can adequately describe.' [4] Glaister was looking backwards to the zenith of modern poison history, the nineteenth century, where if you hang around long enough you'll hear the fruity, resonant language of professors of forensic science pronouncing from the dock in the latest murder case. Like fog, the adjectival amplifications envelop and conceal.

Overstatement has departed from standard texts in the late twentieth century but it lingers on in the storytelling voices of journalists, authors and film-makers. Every so often the Sunday supplement features a modern poison story and I'm intrigued to find the old phrases, as indestructible as matter, appearing in *New Yorker* journalese. Poisoners are still *diabolical criminals*; their interests are *sordid, morbid,* even *wicked*; they hide *depravity* behind smart clothes and an open smile.

The underlying narrative in all poison murder stories is a betrayal of trust. A victim is deceived into swallowing (usually) a lethal substance concealed inside a benign carrier like food or drink. Not only are the consequences life-threatening but a basic trust dynamic we all share — the nurturing mother ideal — is callously disabled. We need to trust those who feed us, those who put medicines in our sick mouths. The hand on the spoon has no business offering us deadly additives. So it is no surprise that the seed-bed for the language of poison narratives is found in the Mother Archetype. Carl Jung wrote: 'On the negative side, the Mother Archetype may connote anything secret, hidden, dark; the abyss, the world of the dead, anything that devours, seduces and poisons, that is terrifying and inescapable like fate.'[5]

To understand the psychology behind our reactions to poison narratives, we need to descend into the realm of the dead, passing along the way the dark, hidden chambers that belong to the Mother Archetype. The myths that embody archetypes tell us something about our morality, our faith, our history; poison stories ground a particular archetype, the shaman/healer, in everyday reality.

The audacity of Cleopatra's story is that she brings all the subterfuge, double dealing and secret calculations of the poisoner out into the open. The secrecy issue is cast aside in the rush to meet other imperatives. Like Psyche holding the lamp to sleeping Eros's face,[6] Cleopatra wants to see the worst and make her choices. We can only imagine, or reconstruct, the machinations of a poisoner before the deed. Data about dosage and likely outcomes might be available but at the critical stage things can and do go wrong. Cleopatra's effrontery (in a narrative sense) turns the magic on its head where we can see the magician's lightning fast fingers at work.

Cleopatra, woman and queen, paid close attention to the iconography of her death scene. The exposed breast, the phallic snake, the dying that looks like post-coital sleep, eroticise her posthumous image in ways that remind us of the living woman who animated war and leadership with sex. The costuming, staging, and carefully positioned artefacts were a rich source of narrative embellishment for the myth makers. Octavius's officer asks of the last dying maid: 'Charmian, was this well done?' To which Charmian replies: 'Perfectly well and worthy of a descendant of the Kings of Egypt.'

Images of a daughter of the sun conducting science experiments with her physician clutter up the clean lines of the narrative sweep: Cleopatra's suicide is usually told without the distractions of hemlock and laurel water and convulsing prisoners. As I read further I found myself drawn in a perverse way to the surrounding detail and away from the central theatre of the fangs. I struggled with the image of the prisoners served up their cups of poison and told to drink. Drink *or what?* I wondered. Drink or be killed? Did one of them say 'No, I refuse' and cop a clout over the ears for disobedience yet still go on refusing until he was either held down and force-fed, or dragged away for execution? I wondered too about Cleopatra's doctor, the expert in matters of life and death who had to borrow knowledge from 'certain women of the city' yet couldn't produce a poison perfect enough for a queen *in extremis.*

There is a familiar narrative, as opposed to historical, integrity in 'easy death' stories. The soft swoon, the quiet slipping away, the pale but beautiful corpse, are all recognisable motifs from fables and myths. In fairytales, characters just die, or they *go to sleep,* a pretence that is endlessly recycled in novels and films. By idealising the death-bed, the living are spared the untidy mechanics of mortal departure, the gaze is protected, the human resistance to engage head-on with extinction is left intact. Of course, anyone who has been at a real death-bed knows that easy death is a rare thing and some passages through the dark gate are unspeakably grim, yet we retain a fondness for 'smooth passage' fables that put a framework of order around our daily experiences of chaos. What is plain to see about death can be acknowledged, even accepted, yet the need to trade in euphemisms — and pay attendants to disguise the mess dying leaves — persists.

The longing to be beautiful in death seems to affect personalities with a vain eye on posterity as much as those who wrestle with fear of the final moment . A few days before her suicide Eva Braun told Frau Junge, Hitler's secretary, 'I want to be a pretty corpse. I'm going to take poison. I wonder whether it hurts. I am so afraid to suffer for a long time.'[7] After swallowing her cyanide, she was reportedly found lying as though asleep with a smile on her lips. Adolf Hitler shot himself immediately after swallowing his dose, a precaution that suggests a determination for self-annihilation over narcissistic scruples.

Poet Anne Sexton told her psychiatrist in 1964, 'I worry about the minutes before you die, that fear of death . . . I'd do anything to escape that fear. I'm so fascinated with Sylvia's [Plath] death: the idea of dying perfect, certainly not mutilated . . . Sleeping Beauty remained perfect.'[8] Sexton at that time believed a combination of alcohol and pills would deliver her a fairy-tale ending; ten years later when she composed her real death scene, she chose vodka and exhaust fumes from her old red Cougar — entertaining those dreaded last moments wrapped in her mother's fur coat, listening to the radio.

>-+-•>-•-○-•<•-+-<

Poison is rarely if ever perfect. Toxic literature in all its vastness and chilling detail records very few compositionally perfect endings. Poison emerges as a fickle, slippery agent with a nasty habit of departing from the poisoner's script. Cobra bites are painful, and death — if it comes — can take up to half an hour. Survivors have told us what it feels like to be bitten; drooping eyelids, difficulty in swallowing, saliva running down the chin, dizziness, nausea, and a nervous pulse. (Some report feeling very little of anything, even pain.) The ones who die succumb to heart failure or suffocation following on lung collapse. Shakespeare conveys the struggle in this last glimpse of Charmian, Cleopatra's maid:

> ' . . . tremblingly she stood.
> And on the sudden dropped.'

When Cleopatra's physician cautioned against the asp ('there is much that is *not* pleasing about a corpse that is viper-bitten') he meant the swelling, skin discolouration, internal bleeding, organ failure, shock, spasms and collapse caused by *Vipera aspis*.

If it could be achieved, would a perfect death be like Cleopatra's? Perfection is a relative term — a perfect act lacks nothing, is utter, unmitigated, and completely suited to the purpose or occasion. The notion of perfection in relation to death is duplicitous. In Plutarch's account, (written at a distance of 200 years after the events), Cleopatra was in a dismal state of grief for Mark Antony not long before she committed suicide. Her thirty-nine-year-old body, far from being unblemished, resembled a battlefield: she was '. . . marvellously disfigured: both for that she had martyred all her face with her nails; . . . her eyes were sunk into her head with continual blubbering; and moreover they might see the most part of her stomach torn asunder.' The leap from this picture to the picture of contented deathly repose on her golden couch was achieved by her maids, Iras and Charmian, who laboured like mortuary make-up artists to patch up and fill out the cracks with creams and rouge. The new perspective diverts the gaze away from the imperfect body — as though it were not permitted to exist — to the idealised terrain of sleeping (but dead) beauty.

In 'Ode to a Nightingale', written not long after the slow horrible passing of his brother from tuberculosis, John Keats wrote: 'I have been half in love with easeful Death.' Most of us, given half a chance, would go for the easy way out. 'To cease upon the midnight with no pain' seems infinitely preferable to suffering and fear. But how do you do it? Keats had studied drugs and poisons at apothecary school; he gives some thought to hemlock, opium, and alcohol.

The chemical means of getting to that quiet, painless death state by a single agent has not yet to my knowledge been found, though many have looked. Dr Kevorkian's assisted suicide machine is too specialised (not to mention illegal) for the lay person, and the advice given by 'self-deliverance' societies involves multiple steps, doctor-shopping and secretive stock-piling. Killing yourself with drugs or poison is not easy or painless; it's mostly a messy, unromantic business often involving blood, spasms, bad smells, leakages, and frightening things like fighting for breath. Hemlock does bring on a 'drowsy numbness' but it also makes you blind, sick, giddy, deathly cold and breathless. Opium makes you vomit. Alcohol draws out all the demons of disorientation to push you over the edge into ethanol poisoning.

Does a perfect poison exist? What would it look like and where would you get it? Fiction writers have always believed in its existence. Dorothy L. Sayers wrote of one which could 'slay in a split second and defy the skill of an analyst'.[9] This is a popular idea. If we only looked hard enough, far enough, if we could only manage the technology or track down the native arrow poisoners and steal their recipes, we'd find an easy ticket to an easy death.

To find out why Cleopatra's prisoner turned blue at the mouth after his laurel water, the interested reader can look in any standard twentieth-century text on Hydro-cyanic Acid. 'Cyanide produces a cytotoxic anoxia by reversibly inhibiting cellular oxidising enzymes containing iron in the ferric state', which freely translates to a deadly depletion of cell oxygen. Is this sort of detailed exposition useful in defining and understanding poison, or do we learn more from Blake's jingle in 'Auguries of Innocence': 'The strongest poison ever known/came from Caesar's laurel crown'? Inevitably, the search for wholeness and faithfulness must encompass, but extend beyond, what is written by experts.

Many have looked for perfection in death (today we

have how-to books and websites dedicated to the subject), yet none is easy, none can be gracefully achieved without help (without those faithful maids to clean up the mess). Do our blunderings have something to do with the dark forces conjured up at death's door?

Chapter Six

There's not much in dying

AUDACITY IN THE FACE of stern underworld gods was given the French Realist treatment when Gustave Flaubert wrote a graphic account of death by arsenic in his nineteenth-century novel *Madame Bovary*. If Emma Bovary had swallowed cyanide instead of arsenic, her death scene would have been a paragraph long. Sip, gasp, clutch, sigh and straight into death spasm. If life had really been a fairy tale, she might have eaten a deep and subtle poison (unspecified like the ones in Grimm) and, like Snow White, have simply fainted — falling gracefully into a state of suspended animation, still red-cheeked, still painfully beautiful, fresh and life-like. But Emma chose arsenic.

'There's not much in dying,' she says to herself after cramming white powder into her mouth. She is lying down and now waits for what she imagines will be her blissful passing from life's woes. Emma poisoned herself. She wasn't tricked into eating an apple by a disguised queen. Her life's drama was self-directed to the end. The first hint of misery comes as an inky, acrid taste in her mouth, accompanied by an unquenchable thirst. Ten pages later, her tongue is protruding from a gaping mouth, her eyes are rolling back, and, then, mercifully, she convulses and dies. But there is more. When her body is dressed at the laying out, a line of dark, horrible fluid leaks from her mouth.

Her husband, and various cold-blooded witnesses, including the pharmacist from whom she stole the arsenic, gather at her death-bed. In death she is seen to be humiliated by the instrument of death, but is beyond knowing or caring who profits from her hubris. Her death throes might be interpreted as a kind of coming to awareness that death is as serious as life, albeit too late. We are excluded from Emma's mind during her final ordeal. We, like Emma, are forced out of the cerebral into the physical.

Seized by grief, Charles Bovary tries to make amends. How can he show his love? He orders an elaborate laying-out. Emma is to be dressed as a bride with a wreath encircling her long, fanned-out hair. He orders three coffins to lock her away like a Russian doll. He orders a green velvet throwover. The witnesses are overwhelmed with wonder at a gesture so remote from everyday life. The expense! The superfluity!

What is the fictional Emma's death all about? Why couldn't she fade into nothingness and expire with her fantasies intact? The idea of a smooth transition from life to death has always appealed to thinkers, writers, poisoners and suicides. Flaubert, the son of a doctor, was, however, clear in his own mind about romantic notions of death. They were part of the fiction of middle-class life. Emma beats her breast for her lost love, Léon, in between reading love stories, burning incense, and letting down her hair like Rapunzel from the wide-flung window, all as a prelude to ending it with poison. If Emma had written the script herself, her life story would have ended right there on the bed, with her body prettily arranged, her face pale but serene and her loved ones weeping silently. Flaubert's relentless reality, however, continues to hold the camera on the witnesses, who begin to quarrel, before growing bored with the drawn-out proceedings.

In the fairy tale, Snow White's pseudo-death granted the dwarfs permission to enter into rituals. In traditional manner, they washed her body in water and wine, then sat down to weep for three days and three nights, by which time putrefaction ought to have, *but did not*, set in.

Emma Bovary's body began to decay with alarming authenticity. Monsieur Homais, the chemist, fetched camphor, benjamin, aromatic herbs and chlorine water to 'sweeten the air'. After some time of sitting vigil, the chemist and the cleric took snuff to clear their heads. Dogs began to howl. 'People say they scent the dead,' the priest remarked. When the sitters fell asleep, Charles paid his last respects to his wife. Emma, it seemed, was disappearing before his eyes. He couldn't resist taking a final look. His shriek of horror woke the household. When the workmen arrived in the morning, Emma's body was sealed into its three coffins.

The dwarfs baulked at putting Snow White into the black ground. To proceed to the second step in the ritual act, burial, the first step (extinction of mortal life, followed by signs of decomposition) had to be witnessed. Yes, the girl was still and silent; her chest did not rise and her eyes did not open, but could this strange state be death? Death did not look so pretty or smell so sweet. These were men who trained at Nature's college. They loved their house-guest, so they had a glass coffin constructed for her to sleep in, and they settled down to see what would happen next.

Inside her glass coffin, Snow White reposes while time passes and things happen. An owl, a raven and a dove–all symbolic birds — fly down, stay a while then leave. Then, the king's son, the animus figure in the story, wanders into the little grove where she lies. By staying still, and letting time pass, Snow White has achieved a merging of passive and active energies that will help her win the fight against the old queen.

How the moribund Emma might have wished *her* Prince Charming had wandered by. As the realisation of her suffering grew, she cried out 'Oh, God, it's ghastly!' Her fear for self expanded to embrace her child, her husband, was *he* perhaps her saviour? had he been all along? But for Emma, there was no final slotting together of separated parts. The slow poison corroded what was left of her eaten-out life. Charles, her husband, after a display of grief at the funeral, finally grew calm and 'vaguely relieved that it was all over'.

><+>+0+<+>+<

Over the centuries the word 'poison' has acquired a supernatural glow. In the past a poisoner and his potion used to levitate a few feet above the ground, the way serial killers do today. Quite a deal of this mythologising has come about, I believe, because poison is not easy to define. Galen had his ideas in the second century,

Avicenna in the eleventh, and in the sixteenth Paracelsus wrote, 'All substances are poisons; there is none which is not a poison. The right dose differentiates a poison and a remedy.' Literally this meant that water could be called a poison if, for instance, a person with heat stroke was given ice cold liquid and died of shock. Similarly, ground glass mixed with sugar (a once popular form of revenge) fitted the definition by killing through mechanical abrasion. The idea is engrossing, and in its time provoked debate and opinion, but it hasn't survived the forward march of forensic science, nor has it dented the popular belief that crimes of poisoning involve sinister characters practising dark arts. Twentieth-century toxicology defines poison as 'any substance which destroys life or impairs health by reason of an *inherent* deleterious effect.' But even modern definitions have a bit each way. Sub-classes appear in some texts: corrosives, irritants, narcotics, narcotico-irritants.

'Poison', from the Latin root *potio* (a draught or drink) is a generic term. Its generous reach gathers in armsful of plants, herds of animals, quarries of minerals and skies of gaseous elements. Nearly all the members of its vast cast can be described and named, even those far-flung rarities that haven't yet met science's gaze. But is naming the same as knowing? In the world of physical reality, how do we recognise the poisonous principle when it has no mug shot, no *Wanted* poster on the post office wall? It might be a white powder, a green paint, a black tincture, a yellow bean. It might glow quietly on the forest floor, it might trail tentacles in a warm Pacific sea.

In general, most of us know what's bad for us. We know about toadstools, oleanders, rhubarb leaves and belladonna berries. We've heard about spiked drinks at parties. The bit we don't know is how to detect cyanide in laurel bushes or peach kernels, how to tell if the kelp in our herbal tablets was harvested in arsenic-rich seawater. Every human orifice is capable of receiving poison, as is each pore, each lesion, each break in the skin. We have to rely on inbuilt early warning systems to scrutinise what's out there. Our noses twitch when smoke invades the bedroom, our skin prickles under spidery footsteps. Our ears hear the rustle in the undergrowth and our mouths spit out methanol labelled 'lemonade'. Most of the second-by-second trawling of our visual sweep is benign and then, miraculously it seems, a shard of glass shows itself in the beach sand and we're saved from pain. The inference we draw from children eating mothballs and soap or old men feeding on weevilly oats is that sense faculties need to be trained in youth and maintained in good working order through life.

The profile of the typical poisoner is as elusive as poison itself. The consistency, if any, lies in the deed. If we are to believe the written record, poisoners employ subtleties under a victim's nose. They smile while adding Circe's 'vile pinch'. The potion-maker enjoys the luxury of being able to separate act and consequences by hours, days, weeks, even years, secure in the knowledge that death by degrees of pain and wasting can be laid at the door of many kinds of mischief, including organic disease. Many poisoners have been women and the act of poisoning has been described as an extension of food preparation, as if it belonged on a continuum of feminine employment and therefore required only domestic dexterity.

To ask 'What is a poison?' in this new millennium may not be the question. 'Why use poison?' might be the swipe-card that opens the locked door. What is to be understood by the persistence of old-fashioned backyard methods? An Australian woman who recently put oleander extract in her husband's coffee, a little at a time over two months, succeeded only in making him ill and suspicious. Her clumsiness in misjudging the dose might signal any number of things: her inexperience, her fear of going too far too soon — but what it signals to the general public who followed her arrest and trial with great interest is the reappearance of a quaint practice from an almost but not quite forgotten time when our elders used to warn us about those dusty trees over by the fence.

In trying to pin down poison does one veer away from the particular and go into generalities? The sense faculties for instance? Looking will pick up colour, but colour can't be relied on: it varies. Arsenic is red, yellow, green or white depending on its chemical bedfellow. There are red berries we can give our children to eat and red berries which cramp and kill. Similarly, our fingers aren't wired to discriminate the poisonous from the innocuous: many poisonous vessels like cobra skin, belladonna plaster, strychnine nut, mercury drops, opium calyx, almond kernel and the velvety death-cup mushroom are sumptuous to touch. And what of poison's smell? Prussic acid smells almond sweet, hemlock smells like a family of mice, devil's dung smells like garlic, oleander smells like chocolate, and arsenic in cocoa smells like supper on a cold night.

So where is the poison principle if it's not in the shape, colour, feel or smell of a thing?

Chapter Seven

The leap from chaos

IT MIGHT BE NECESSARY to detour around the libraries and lecture halls and pose our question another way, beginning further back than what has been written or said about the great slayer, poison.

We might posit a creation pathway, before time and place, before even shape and form, somewhere in the roaring reaches of outerspace, eons ago, when the seeds of what will eventually become earthly poisons might be seen glittering in the churning shapelessness of flying debris and atomic jumble.

To begin this way is to imply not pre-destination, but its opposite, serendipity. These glittering seeds might spin forever in the great washing machine of unformed space, or something might happen to bring them to earth.

What if something *does* happen. A solid earth grows from this undifferentiated embryo and the poison seeds germinate. Gold, silver and copper thread their coils through the substratum. Arsenic, antimony, mercury, thallium, and a hundred other pure elements take up residence, alone or with partners, in the new neighbourhood.

In this inanimate world there is no good and evil, no demons or angels, only a wasteland rotating at a fixed speed around the sun.

After a billion years the oceans form. Earth elements meet water elements. Old bonds are broken, new ones negotiated.

But a dead atmosphere clings to this earth. Nitrogen, hydrogen and their pungent child ammonia rage at the solid interface but can't get in. A new arrangement is needed: atmosphere, soil, water and fire engage in a battle that propels oxygen and carbon out of hiding and into the open.

A billion years follow, then a smear of green algae stains the water.

Life.

Once it starts, there's no stopping it.

In the seas, tiny single-celled algae manufacture a quaternary base as neurotoxic as strychnine. Higher up the food chain mussels and molluscs concentrate the poison as they feed. In the preCambrian seas, jellyfish like Chironex grow stinging tentacles.

At the last minute, humans join the progression of animated feeders, and it is at this moment, sometime, somewhere, that the leap from chaos to design is made.

Man and poison meet. It might be a chance, fatal encounter with the wrong berries. It might be fish stunned by a vine blown into the river. It might be an arrow smeared with pulp that changes the way man hunts.

But what begins innocently enough will never be artless again.

Chapter Eight

Myth and evidence

WORLDWIDE, MORE THAN 13 million natural and synthetic chemicals have been identified but fewer than 3000 are what we know as poisons. By the 1970s, death from all forms of poisoning in Britain amounted to 4000 per annum, or 15 per million of population.

Relative to other agents, poison is too slow and too hard to get. Like sweet spirits of nitre for fainting ladies, murder by tainted substances is passing into history. Yet by some sort of inverse law, public interest has remained vigorous. Poison stories still sell newspapers.

'Did you read about Swango?' friends asked one morning in 1998, passing the pages of the Sunday supplement around the table. Michael Swango is an American doctor who poisoned his patients and attempted to poison his friends and fiancée for 'no apparent motive'. 'I think,' an Illinois judge is quoted as saying, 'he wanted to take them to the edge of death. They were like a lab experiment.'

George Orwell described the fascination of the 'good

read' in his 1946 essay, 'Decline of the English Murder'.[10] 'Our great period in murder, our Elizabethan period, so to speak, seems to have been between roughly 1850 and 1925 . . . The background of all these crimes, except Neill Cream's, was essentially domestic; of the twelve victims, seven were either the wife or husband of the murderer.' Orwell lamented the passing of the 'old domestic poisoning dramas, product of a stable society where the all-prevailing hypocrisy did at least ensure that crimes as serious as murder should have strong emotions behind them.'

Orwell's post-1920s murder is increasingly anonymous and technological. Retractable knife blades, repeating, fast-loading guns, silencers, AK-47s have blown away the chemical competition. Killing is quicker, more certain, and regrettably, if we agree with Orwell, far less interesting. Domestic crimes involving people who know each other well speak to the reader of passion. Road rage shootings, Oklahoma bombings and corpses at McDonald's speak only of external chaos and futility.

The Swango (and more recently Shipman) murdering-doctor stories fall short of Orwell's high standards. Coleridge's phrase 'motiveless malignity' which referred originally to Iago's crimes, and pre-dates the potent terminology of forensic psychology, is as close as most of us will get in thinking about how to label irrationality. For sense and meaning we turn to fables. Emma Jung wrote, 'when a fairy tale is told, the healing factor within it acts on whoever has taken an interest in it and allowed himself to be moved by it in such a way that through his participation he will be brought into connection with an archetypal form of the situation and by this means enabled to put himself "into order".'[11]

Deeper levels of sense and meaning are available to us in the often confusing and overlapping versions of world mythology. In this rich legacy of floating talk it's the message of the symbols, not the literal rendering of a story, that help us break through to an understanding of the way things are. While we enjoy Jason and Medea at the level of an adventure story, it's also possible to read Medea the witch as Jason's anima figure, the inner feminine energy whose occult powers solve what is, for Jason's ego, otherwise insoluble. When Jason later rejects Medea (and therefore the contribution of his feminine knowing) she retaliates with murder and the laying of curses, and Jason withers and finally dies in an atmosphere of decay.

The extraneous details of great (and small) stories are wisely put to one side by writers whose job it is to tell their tales well. The trimmed-off pieces live on as ephemera, scraps of dialogue, repetitions, clues that led nowhere in particular, pointless interviews and confessions and embarrassing examples of the writer's struggle to make the facts fit the picture. In the canon of poison literature these homeless bits of paper are shuffled around until they find friendly accommodation, usually in archives of some kind, or they lodge as unfinished business in the living memories of the aggrieved. In these tailings, these winnowed-out, edited and aborted leftovers lie, I believe, some of the truths about the way we live and die.

Chapter Nine

Lethal greens

SIFTING THROUGH POISON lore I was struck by the number of references to the colour green. The links were obvious or implied or creatively oblique but they were there. If I was looking for the poison principle this might be the solid clue amongst thousands that leads me to the murderer's door.

Look at the company it keeps. Green is the colour of new leaves, forests, growing grass, unseasoned wood, emeralds, money, jealousy, gullibility, fear, witches and sea dragons. It vibrates at the centre of the electromagnetic spectrum and is the *benedicta viriditus,* the blessed green of alchemy. Meditations on green soothe the heart, blood, lungs and immune system. Thinking green light into the heart chakra calms our throbbing engine room, steadies its pumping beat. Emerald green is associated with Hermes, the trickster god of the Greeks, and with the heavenly planet Mercury, ruler of the constellation Gemini. In Hindu gem therapy, an emerald stone embodies the blessed green of healing.[12] Worn on the

Mercury, or little, finger emerald imparts control over the nervous system, liver, vocal cords, tongue and brain. It animates wearers who are lazy. It profits people born under the sign of Gemini or Virgo. Crushed into powder and taken orally, emerald cures snake bite. Poets make striking, unthought-of connections to green. In 'The Garden' Andrew Marvell, inspired by a term in rural Yorkshire, wrote rapturously about verdant nature.

> Mean while the Mind, from pleasure less,
> Withdraws into its happiness:
> The Mind, that Ocean where each kind
> Does streight its own resemblance find;
> Yet it creates, transcending these,
> Far other Worlds, and other Seas;
> Annihilating all that's made
> To a green Thought in a green shade!

Omar Khayyám tore himself away from the delights of wine long enough to notice the appeal of a bit of chlorophyll:

> And this delightful Herb whose tender Green,
> Fledges the River's lip on which we lean —
> Ah; lean upon it lightly! for who knows
> From what once lovely Lip it springs unseen!

But much that is green by nature, complexion, or chemistry is *not* blessed, sacred or even safe. The benign properties of green thoughts, green gems, green shoots argue a corresponding set of malign, not so nice properties. Toads, hemlock and some arsenics are green. My grandfather's murdering tonic was green. We inhabit a world of duality and opposites where every coin has a flip-side, every day is followed by a night. Green colour might be a double-dealer; the fabled emerald city might have a dirty, downtown heart — there are caveats that the investigator must attend to.

Professor Glaister of Glasgow had something to say about green colouring in his classic poison text of 1954. It seems that green-tinted confectionery and cake icing put the fear of poison into the minds of rural Scottish folk. They 'stupidly' regarded green as an enchanted colour and would not put it in their stomachs. Speakers at a confectioners' trade convention quoted lost business when they put out green sweets. The Scots refused to buy the 'arsenical-looking things'. One baker had to eat his green-iced cakes in front of customers (like a poison-taster at the king's table) to clinch the sale. Professor Glaister is scathing. 'It is astonishing why the idea of a green colour in confections should suggest the stupid impression that arsenic is present, when arsenic in its common form is white.'

True, the common form is white. But people have long memories. The green form of arsenic, copper arsenite, *has* been used criminally in food, in a blancmange eaten by a gentleman who died at a public dinner; in wallpaper that children who didn't know any better have licked. Arsenic stories belong to the next chapter but, in a green mood, I want to tell you about a lady and her ballgown and suggest other reasons for these nineteenth-century collywobbles about colour.

It is 1862. At the London ball season crinolines are the rage. A young lady is wearing twenty yards of green *tarlatane*. Silk would have pleased her more but tarlatane is a passable substitute since it can be made to shine in a silky way. The tarlatane comes from a factory in Leipzig where they have the knack of laying a paste of starch and copper arsenite (called Schweinfurt Green) onto cotton. When the paste is dried and polished, the cloth dazzles like an emerald.

Fully rigged for dancing in matching gown, headdress, fan and shoes, the young lady is carrying enough arsenic on board to kill everyone in the room. If she dances till the late hours in a jostle of overheated bodies, thousands of lethal green particles, loosened from the paste on her dress, will lift and spin in the whirls and eddies of a room shut tight against the damp air. When she raises her fan, which has lain between times in the folds of her gown, or when her dress is admired and the folds extended, she will dust her partners with green death.

By midnight her head is aching so much she has to leave, pleading exhaustion. Her face is pale. She barely makes it home and undresses before a spasm of nausea tenses her stomach. She needs to urinate every five minutes. She loosens her drawstring and lies down. Suddenly she can't breathe.

She sits up, coughs, clutches her throat. Then she vomits. Her maid finds her slumped over the commode pot, her bowels emptying watery stools, her face a deathly white.

Is it cholera? typhoid? So many young officers are returning from Algiers, Morocco, the East, bringing foreign, purging contagions into the ballrooms of London. She can hardly speak now. Her throat is burning and raw. Her hands and feet are icy cold, the dreadful machinery of her body is trying to expel all its fluids even when it seems there is nothing left.

The doctor arrives at 3 a.m. Diagnosed with choleric dysentery, she is taken to the infectious diseases hospital where she dies the following day of liver failure exacerbated by dehydration.

Stories like this come by way of the *British Medical Journal* which devoted space in its correspondence pages of 1862 to the ubiquitous presence of arsenic in daily life, citing ballgowns, wallpaper, flypapers, tints, dyes, and paint. Dr Letheby, author and commentator on the problem, advises ladies (in his misogynous way) to carry a 'small phial of ammonia when shopping instead of the usual scent phial. The mere touch of the wet stopper on the suspicious green would betray the arsenical poison and settle the business immediately.' He is referring to a container of strong liquid ammonia which will turn blue in the presence of copper. 'Copper is rarely if ever present in these tissues without arsenic being present, the compound being copper arsenite.'

Deaths from wallpaper arsenic, however, were more likely to touch the lives of country citizens than were ballgown tragedies. The 1862 *British Medical Journal* reports an inquest into the deaths of four children in one family. Before dying they suffered sore throats and 'extreme general prostration' from daily contact with unglazed paper later found to contain three grains of green arsenic per square foot. Two children from another family died after as little contact as one hour per day playing in their father's library. Dr Blyth, writing in 1884, notes:

> Recently I examined a wall-paper, which, on
> analysis, yielded arsenic equal to 8 grains of
> arsenious acid in every square foot; it was a portion

of a wall-paper from the bedroom of a gentleman,
who had for many months suffered from an obscure
malady, which had baffled the eminent men whom
he had consulted. The paper in question had a few
green leaves on a quiet drab coloured background.
On the discovery of its arsenical properties he
removed to another bedroom, with immediate and
marked improvement in health.[13]

Dr Glaister's bullying argument about the innocence
of green no doubt impressed his medical peers, but in the
houses where children died from inhaling or swallowing
green arsenic dust the lay person faced his eloquence
with a silent and knowing rebuttal.

Cora Crippen, who was to die of poisoning in 1910, had
no doubts about green. 'Having seen a green wallpaper in
the drawing-room of her friend's house, Mrs Crippen
expressed herself shocked and said: ' 'Green paper! You'll
have bad luck as sure as fate. When I have a house I won't
have green in it. It shall be pink right away through luck,'
and apparently nearly all the rooms in Hilldrop Crescent
were decorated in this propitious colour."[14]

'Good' green drugs persist into the late twentieth
century like artefacts of a bygone era. Some, like digoxin,
begin life in plant form, but the white or blue tablets that
heart patients shake out of their bottles each morning
are as far from foxglove as this page is from a tree. 'Bad'
green plants on the other hand never go away. They have
a persistence that is rooted in the human need to transcend or annihilate certain realities; they stand on the
boundaries between the tamed and the wild. In fables, as
in Eden, there are sensible cautions against eating certain
vegetation. Bad plants are really hidden passageways into

other worlds; if you eat them you open yourself to being spirited across the known-world boundaries to who knows where? — a short trip to a fantasy destination or a frightening one-way slide into a claustrophobic cave. In these alien territories, if you eat the fruits of the realm you may be stuck there forever — unless a spell-breaker arrives.

The desire to transcend the known-world mediocrity finds its own wilful expression. The trickster leaves plenty of keys lying around. Doors fly open. We dance, trance, fast, meditate, get into pain, or mess up brain chemistry. Boundary drugs pull us out of the Kantian mind-set that names and defines; they drag us to the lip of a new abyss. Sometimes we float out and above, sometimes we're left blinking in the darkness below.

>━━◆◇◆━━<

The poisonous potential of certain plants was known from earliest hunter–gatherer times, was refined in the successive cultures of Mesopotamia, Egypt, Greece and Rome, reached a sort of apotheosis in the Middle Ages and a revival in the Renaissance. Once upon a time we could find out which plants were lethal, or at least overwhelming, from witches. If you wanted a brew to kill a husband, or impassion a lover, you went to see the old woman who knew her botany. Witches had a take-it-or-leave-it attitude: their hoydenish behaviour bartered with the superficial layers of the buyer's conscience. Most of the 'greens' we look at in this chapter were stock in trade to practising witches, herbalists and apothecaries of the Middle Ages. The complete list would fill the page: aconite, belladonna, buckthorn, buttercup, cannabis, cinquefoil, wild cherry, darnel, poison dogwood, dumbcane, English holly, foxglove, heliotrope, hellebore, hemlock, henbane, laburnum, lantana, larkspur, lobelia, mandrake, mistletoe, monkshood, morning glory, mountain laurel, nightshade, oleander, opium, poinciana, poplar leaves, raspberry, rhubarb, rue, scotchbroom, smallage, spurge, tansy, thorn apples, verbena, and wormwood — all these are an alphabetical beginning only.

Lethal green herbage grows alongside the unremarkable varieties without benefit of labels. Witches knew the ones to pick. They made green ointments to smear on their bodies before flight. These rubs, described by Duerr, usually consisted of raw plant (leaf, stem, flowers, sometimes root) mixed with grease, or if you subscribe to the myth, the fat boiled off the exhumed corpses of babies.[15] Old World herbals called these plants 'hexing herbs' (today we call them *psycho-active*, not necessarily a semantic improvement). What they have in common is an hallucinogenic heart or 'tropane', which acts on the brain to create altered states of awareness. It has never been difficult or esoteric to find these 'tickets to ride'. English country gardens will yield ninety-five per cent of them, the rest and many more of a long list grow in Asia. Hans Peter Duerr has exhaustively studied witches' salves and wisely refuses to dish out recipes more specific than tansy, hellebore and wild ginger fried in butter with an egg.

The nightshade family contains the toxic quartet belladonna, datura, henbane and mandrake. In the Old World, belladonna strayed into the night-time business of witches, today it's a tame little thing grown by suppliers to drug companies. When I was a student we

used to add belladonna tincture to magnesium trisili-
cate mixture, being careful not to show any dark green
particles in the finished product. Before Tagamet, Zantac
and Losec, this mixture was a standard remedy for
stomach upsets. It survives today as its derivative,
atropine, in antispasmodic tablets like Atrobel,
Buscopan and Donnatabs. *Atropa belladonna*, or deadly
nightshade, is a herb with shiny purple-black berries
found in the woods of England, Europe, Africa and Asia.
The old custom of putting a few drops of its juice into
young girls' eyes to achieve the startled fawn look that
was said to enhance beauty (hence the descriptive *bella
donna*) prevails today as the rather more mundane
Atropine eye drops used by eye doctors to enlarge the
pupil before examination.

Belladonna poisonings, accidental and otherwise, are
inevitable. In 1884 a French doctor was gaoled for eight
years for sending thrushes poisoned with belladonna
alkaloid to a rival doctor whose wife and servant ate
them. Forbes reports a case from the Old Bailey where
tarts made from belladonna berries killed a husband and
wife.[16] (The pedlar who sold them claimed they were
huckleberries.) More recently, a three-year-old Queens-
land girl, who had chewed some green (unripe) berries,
frothed at the mouth and lapsed into a coma. In London,
an elderly woman became comatose after applying
belladonna plasters to raw skin on her back.

Pliny the Elder describes (in AD 23) patients chewing
belladonna as a form of anaesthetic before and during
painful operations. Culpeper can't find a good word for it:
'this nightshade bears a very bad character as being of a
poisonous nature.'[17]

Datura stramonium, called by many names including

thornapple and Jimson weed, is a solanaceous plant with distinctive trumpet-like flowers. A salmon-coloured variety grows in my garden. Sometimes I look at it and marvel at the audacity of people who experiment with the breed. It has a long history. Priests of Apollo at Delphi chewed the leaves before making prophecies. Indian thieves doped their victims with it 'in the service of Kali'. A common Renaissance poison called 'black madness' was a mixture of thornapple and henbane. There are reports of its use in puberty rites in Mozambique; of Zuni Indian priests chewing the roots to commune with spirits of the dead.

Datura stramonium has been a popular intoxicant by what was called the 'psychedelic set' since the 1960s. Of 212 cases of datura intoxication surveyed by a researcher, there were five deaths. When you mixed datura with marihuana you got a mind-blowing blend called 'green dragon'. If you had no datura you could go to the chemist and buy Potter's Asthma Remedy cigarettes. This datura-containing preparation consisted of fifty per cent stramonium, twenty-five per cent potassium nitrate (to make it burn), belladonna, and grindelia or tobacco as a bulking agent. In 1976, a nineteen-year-old British man who attended the Knebworth music festival unrolled eighteen Potter's Cigarettes, poured on hot water, let it soak, then drank the liquid. Three hours later some friends found him in a state of acute disorientation and hyper-agitation. Afraid, they took him to his parents, who called the family doctor. Another five hours later he was still confused, manic, and hallucinating. He had fixed, widely dilated pupils from the belladonna, a pulse rate of 110 beats per minute (normal is 60–80) and elevated blood pressure. After a restless, anxious night in hospital he recovered with no recall of what had gone before. The sleep caused by datura may last up to twenty days.

The third member of the toxic potato family quartet is henbane, *Hyoscyamus niger*, a sticky, hairy and foul-smelling inhabitant of wastelands and rubbish dumps — the bane (death) of hens but the boon of hogs who relish the foetid weed. Shakespeare, who spent time in Bucklersbury Markets picking up useful information from plant sellers, chose henbane as the poison for Hamlet's father (Act I, Scene V):

> Sleeping within mine orchard,
> My custom always in the afternoon,
> Upon my secure hour thy uncle stole
> With juice of cursèd hebenon in a vial,
> And in the porches of mine ears did pour
> The leperous distilment . . .

So much has been written about mandrake it needs only a cursory mention here. The plant's forked roots, once likened to human form, were said to expel demons. In Pliny's time, patients chewed the root as a simple painkiller. Before surgery a patient was given equal parts of opium, mandrake and henbane pounded and mixed with water, dipped in a rag and held under the nostrils. The legendary powers of the plant to strike a man dumb or dead were trounced by Gerard in the late sixteenth century: 'There hath beene many ridiculous tales brought up of this plant, whether of old wives or some runnagate Surgeons or Physickemongers . . . you shall henceforth cast these tales out of your bookes and memory . . . they are all and everie part of them false and most untrue.'[18] The tale about it being an aphrodisiac limps on, however. The American FDA, citing mandrake, 'agreed with the panel [convened in 1989] that none of these products which claim to arouse or increase sexual desire or improve sexual performance are safe or effective.'

><—•>—O—<•>—i—<

A greater menace to health is monkshood (also called wolfsbane), the source of the poison aconite. I first took an interest in the awesome powers of this attractive border plant after reading a remarkable case of survival reported in the *Sydney Morning Herald* in 1994, and then reading backwards to Rome and Greece. The contemporary story concerned a thirty-two-year-old visitor to Sydney who bought some tonic for his sore joints from a Chinese herbalist. Thirty minutes after drinking the herbal stew containing aconite the man began to sweat and go numb at the lips and mouth. Soon his hands changed colour (purple-yellow-purple), his forearms felt 'tight', he began to 'spin out', and finally couldn't breathe. Four hours later he was clinically dead. The twenty medical staff who saved his life (with three hours of heart massage, twenty-five electric shocks, five injections of adrenaline and a temporary heart bypass) could hardly believe it when he opened his eyes at the end of it. To bring about the miracle, blood from the right atrium was passed through a filter pump in a heart–lung machine, then oxygenated and returned through a tube in the aorta.

Aconite is a schedule four restricted drug in this country. Prosecutions under the Poisons Act followed swiftly. Looking at the photograph of the man who lived through the double ordeal of the illness and its cure, I was struck by how lucky he was to have had his cup of poison at the end of the twentieth century. How fortunate to have collapsed in a major city within reach of an ambulance and immediate, breathtakingly intricate attention.

Plutarch reports that Mark Antony's soldiers — on short rations — dug up aconite roots, mistaking them for edible parsnips. Juvenal makes his usual cutting comments when satirising a *cause célèbre* involving aconite from Nero's reign:

> How I wish that it *was* all nonsense! But listen to
>   Pontia's
> Too-willing confession: 'I did it, I admit I gave
>   aconite
> To my children. Yes, they were poisoned, that's
>   obvious —
> But *I* was the one who killed them.'
>   'What, you viper,
> *Two at one meal?* The brutality of it! *Two*
> You did away with?'
>   'Indeed; and if there'd been seven
> I'd have polished *them* off, too.'[19]

William Turner in the 1550s called it 'the most hastie poyson'.[20] The herbalist Gerard wrote of its effect — 'so forceable that the herb only thrown before the scorpion or any other venomous beast, causeth them to be without force or strength to hurt, insomuch that they cannot move or stirre until the herbe be taken away.'

There are any number of famous aconite poisonings and more to say in a later chapter. The Lamson case of 1881 is often quoted because Dr Lamson studied under one of the kings of forensic toxicology, Professor Christison, turning what he learnt to his own venal ends. He poisoned his nephew-by-marriage, eighteen-year-old Percy John, with an aconite capsule (or cake, depending on who you read) and was hanged in 1882. Lamson had presumably stored away for future reference Christison's useful claim that aconitine, the poisonous principle of aconite, was untraceable. In the 1870s this was a valid assertion.

This century can afford to take a benign view of the vegetable kingdom. We have rendered up its secrets, catalogued its members, separated out the active ingredients from the dross. Most of the alien photosynthesisers have been tamed, but, flourishing under the noses of the regulators and law-enforcers are forests of untamed, illegal greens whose pedigrees are becoming as intricate as the sniffer dogs who track them. I refer of course to *Cannabis sativa*, one of the most disputatious of green drugs.

Medicinally it's been in use since 2730 BC. Hemp root eased gout, hemp tea helped to stabilise appetite in wasting diseases (today, it's found a niche in AIDS and cancer management). Tinctures were given for uterine haemorrhage, cystitis, urinary infections. Applied to the skin, it was a salve for gunpowder burns, birth pains, colic, jaundice, ague, nosebleed and liver obstruction. Culpeper, the pragmatist, rated it 'good to kill worms in man and beast; the juice dropped into the ears kills worms in them, and draws forth earwigs or other living creatures.' But mostly it finds favour for its exhilarating qualities.

Much of what we know of the early history of cannabis comes via Marco Polo writing in 1300; and, prior to the eighteenth century, cannabis eating and smoking was mainly confined to Islamic and Indian regions. When Napoleon's soldiers returned to France from the Levant, cannabis found a conduit into Europe. By the 1840s it was available through apothecaries, ships' crews, and the ever-present black market. In 1843 *Le Club des Hachichins* began to meet monthly in a hotel in Paris. Guests were given a small spoonful of green hashish jam, a cup of Turkish coffee and a divan to recline on. Gérard de Nerval, Théophile Gautier and Charles (*Les Fleurs du Mal*) Baudelaire were regular club-goers, though Baudelaire eventually grew to mistrust the 'chaotic devil' hashish. (Eating hashish never took off in England; the British preferred opium.)

The active principle, tetrahydrocannabinol (THC), has a powerful narcotic action; and a minimal fatal dose was calculated by Chopra and Chopra in 1939 for *charas*, *ganja* and *bhang* respectively of 2, 8, and 10 grams per kilo of body weight, though it is difficult to find any recorded fatalities. Toxicity presents as restlessness and anxiety ('spinning out'), increased heart rate and blood pressure, cold extremities, dilated pupils, injected conjunctival vessels (red eyes), and ataxia. In extreme cases this progresses to respiratory depression, psychotic episodes, and paranoid delusions.

In places where marihuana has been grown legally, it's possible to find relief from the didactic tone of modern debate on its use. The *Illustrated Weekly* of India of 4 July 1937 joyously describes a harvesting ceremony in Uttar Pradesh: 'Frequently the treading is done to the rhythmic beat of the tom-tom, supplemented sometimes by peculiarly shaped wind instruments, to add a touch of gaiety to the work. The beat of the drum, the wailing of the pipes and the weird contortions of some of the treaders all combine to lend a touch of *tamaska* to the scene from which the workers derive as much fun as the onlookers.'

In the early 1970s it was grown legally in a fenced-off and guarded garden at the University of Sydney. Pharmacology students, myself included, watched it grow. Some students, I'm told, broke in and stole the tagged and catalogued specimens that were part of an academic long-term study. In her co-authored book, *A Citizen's Guide to Marihuana in Australia* (1977, 1981) Lorna Cartwright makes a case for tolerance. 'Marihuana is a drug. It is not poisonous. It is not severely toxic . . . and it is nothing like heroin or LSD. We should perhaps regard marihuana, which as far as we know has killed no one, with at least the same indulgence as that traditionally accorded wine, cognac and absinth.'

><+>-०-<+-<

When I was typing up an early draft of this chapter, cold westerly winds were banging over my house on their way to the sea. I was tired of coffee and thought of Indian chai, the spicy milk tea you get in Rajasthan. In an Asian cookbook I looked up a recipe that asked for green tea, cinnamon sticks, cardamom pods, honey, and milk. In the back of the pantry I found a pretty cube-shaped box, provenance unknown. It was unopened, still enclosed in cellophane. All the writing was Chinese except three words, *Green Tea* and *gunpowder*. When I opened the box I got a surprise. 'Is this tea?' I asked my husband. The particles rustled, and each one looked like a small neatly tied parcel about the size of a clove. He didn't know either, and though we both agreed that it didn't smell like tea, look like tea, or behave like tea, I put a good spoonful in water in a pot, added the spices and milk and set it simmering. It stewed as I stood at the kitchen window, watching the trees bend to the hard wind. Satisfied that it sort of resembled the chai of memory, I poured a mug and returned to my study. I'd been typing the section on the man who survived aconite poisoning from Chinese herbal medicine, and sat back from the screen holding my drink, reading the words. After two sips, which tasted good, I put the mug down and began to sort through papers when a curious thing happened in my chest. My heart which had been doing its beating unnoticed suddenly began to thud. A sense of dread attacked me and bent me like the trees outside. I sat perfectly still, rigid, waiting. My heart went into a slow thud-thud-thud and my head began to loosen its hold on my shoulders. I was sure I'd poisoned myself and was going to die. My husband was outside, beyond hearing, fixing fibro loosened by the wind. This is it, I thought.

My heart seemed to have lost its sense of how to beat properly. In the bathroom I stared at my reflection, willing the floating bits of me back into the same body. Time slowed too. Anxiously I opened cupboards looking for something to take; it was fear that was killing me, I was sure of that. I took 5 milligrams of Valium with two full glasses of water, lay down and eventually slept for an hour.

How do I analyse the experience? Do I talk about my fear? how a little knowledge can be more frightening than none? From a scientific and investigative point of view an analysis was negated when I threw the box into the garbage. It's possible the tea was adulterated with some cardiotoxic plant matter, like foxglove, or worse. It's also possible that my seizure was unrelated to the tea at all — but time for testing that hypothesis has passed and can't be recovered. What is left is a memory of my body registering and processing a pathological phenomenon, that is the entry of a toxin, delineated by its own rules and proceeding towards its own conclusion (except that I interrupted the flow with diazepam). Michel Foucault writes of disease as 'hooked onto life itself, feeding on it, and sharing in that "reciprocal commerce of action in which everything follows everything else, everything is connected with everything else, everything is bound together". It is no longer an event or a nature imported from the outside; it is life undergoing modification in an inflected functioning.'[21]

Supposing for a moment that the culprit was foxglove, the source of digoxin, and supposing I had taken enough, I might have met the fate of a seventeen-year-old whose parents took him to a quack in 1826 and asked for something to stop the boy's attacks of giddiness. Jacob Evans prepared a decoction by boiling foxglove in water, reducing the volume and getting the boy to drink five and a half ounces. Apparently three drops would have sufficed. The unfortunate boy was desperately ill within thirty minutes, then fell into a coma and died three hours later. Evans was found not guilty of murder on the grounds that the boy's family had approached *him* for treatment, and not the other way around.

Digoxin increases the force of the cardiac muscle contractions, which is why it is useful for failing hearts. The Birmingham physician William Withering uncovered the secrets of foxglove in the 1770s after studying a local remedy for dropsy. The plant has long tapering thimble-shaped bells that are pink or purple depending on where they grow. The leaves look something like comfrey and have been mistakenly used for comfrey tea. In 1980 a retired Ohio steelworker who kept away from doctors and relied on home remedies drank a cup of foxglove tea believing it to be comfrey. The tea left him weak and nauseated and he began to see a yellow halo around objects. He spent a week in the coronary care unit exhibiting tachycardia and unusual heart rhythms but recovered after treatment. Others have not been so lucky. Boiling the leaves in wine was a common method of bringing on fatal heart events in earlier centuries.

⊱—⊶—O—⊷—⊰

Even among Andrew Marvell's garden of innocent plants — the oak, the laurel, the bay, the daphne, and those fruits, peach, nectarine, melons, apples — we can find cups of poison. Take, for instance, Marvell's laurel from whose green cells laurel water (a cyanide drink) is distilled and used in medicinal as well as murderous ways. In 1781 twenty-year-old Sir Theodosius Boughton died after drinking a draught sent by the local apothecary. The draught was supposed to contain treatment for venereal disease but had apparently been spiked with laurel water by a relative who stood to gain from Boughton's estate.

No one touches hemlock any more and all its stories

are historical. One of the hexing herbs, it is altogether a disagreeable plant with an unpleasant odour. King Lear threads a bit of it over his costume as he embarks on his mad scene, symbolically choosing plants of a lower order to represent the bitter harvest of neglect. Known also as poison parsley and spotted henbane, hemlock's other famous association comes via Socrates, who expired with hemlock on his lips. The fatal dose is 130 milligrams and death can take anywhere from three minutes to three hours. In the early stages there is a short period of stimulation (the agonal period), then languor, drowsiness, peripheral partial paralysis, staggering gait, thick speech, accelerated heart, then failure to breathe. Mistakes still happen. Hunters have died after shooting and eating migratory quail that have fed on wild hemlock in the The Balkans.

>-+-+>-+-O-<+-+-<

I can't leave this section without a brief look at the herb wormwood, still with us today but never, probably, to be the subject of so much notice as when it was the prime ingredient of the green alcoholic drink *absinthe*. When absinthe first arrived in French cafés via Algeria in the nineteenth century it drove half of France mad. Soldiers acquired a taste for the bitter green liquid given to them medicinally to break fevers while on duty in the French colony. (The seventy per cent alcoholic content might have had something to do with its intoxicating power.)

Absinthism became a new disease. An addict suffered the combined deleterious effects of alcohol and irritant herb. Rich and poor, old and young approached the bar for a dosette of green liqueur. To sweeten the taste, sugar and water were added with ceremonial care. 'Drinking slow poison', lining up for 'a thousand green glasses', became a way of life.

Vincent van Gogh painted a still life of a carafe and a glass of absinthe. Toulouse-Lautrec painted van Gogh contemplating a glass of absinthe. Vincent in a letter to his brother Theo attributed his stomach problems to this 'bad wine', which was eventually banned in 1915, then reformulated by Jules Pernod using anise instead of wormwood.

>-+-+>-+-O-+<>-+-<

Lethal consequences might be expected from hexing herbs but not, I thought, from common or garden parsley, until I looked at its toxic profile. Parsley fruits, I read, when dried and extracted with alcohol give up a green oil called 'apiol' which was commonly used on the Continent to induce abortion, a lucrative market in pre-contraceptive pill times. Reading further I found the story of a mother who died of apiol poisoning in Edinburgh in the 1950s. She was twenty-one years old, had two children under two years of age, and was pregnant again. In a desperate attempt to terminate the pregnancy she'd been taking European 'abortion pills' daily for three weeks with, apparently, no obvious results. On the last evening of her life, she filled a douche and syringe with soapy water to manually dislodge the foetus, and during this procedure, lapsed into unconsciousness.

Her husband found her lying face down in the grip of a strange sort of paralysis. The lower half of her body was flaccid and unresponsive while the upper half was tense and flexed. At the hospital doctors attempted unsuccessfully to return her to consciousness. Some hours later she had 'a complete spontaneous abortion' and her high temperature returned to normal.

However with each passing hour the paralysis crept upwards, preceded by irregular twitches and flexings, until she finally died never having regained consciousness. The phenomenon was identified as Landry's Ascending Paralysis caused by ingestion of apiol adulterated with triorthocresyl-phosphate. In an Australian newspaper of the 1920s I read and wondered about

> Parisian female powders, that restore regularity and remove obstructions without fail in 48 hours. No obstruction can resist them and they never leave any bad effects. Price seven shillings and sixpence. Write to The Specialist at a P.O. Box in Melbourne.

>‑＋◆▸‑○‑◂◆＋◃

My last look at lethal greens came via an Australian police journal.[22] Opening the magazine at random I saw a heading, 'The Green Dream' — an article written by a female detective from the Forensic Investigation Unit. 'The *green dream* first came under my notice in 1994, when I attended the scenes of two deaths which I investigated where both persons committed suicide by using this lethal substance.' What substance? I wondered, and turning the page (always a risk when looking through forensic information) came across some disturbing *post mortem* photographs of a man and a young woman. The man was in a chair, his head on his shoulder, with an intravenous drip feeding a bright green luminous liquid into his left foot. His suicide note said, in part, 'Being someone who was not born yesterday, I knew where to go in Kings Cross to get a lethal dose of the green dream.' This was the detective's first encounter with the 'alien substance' and she did her homework. It turned out to be a veterinary euthanasia solution, based on pentobarbitone sodium, or Nembutal, withdrawn from sale for human use in this country but obviously still available through veterinary suppliers (from where, one presumes, it had escaped into the wild). Only a few months later, the same detective attended the death scene of a second green dream suicide, this time a twenty-year-old girl who worked at a Sydney research laboratory and is assumed to have obtained the green dream from her workplace. In a postscript, the detective notes that, by 1998, 'four years have passed and the "green dream" appears to have disappeared into oblivion.'

⊷─┼─◁─○─▷─┼─⊶

Having taken notes on everything that is suspiciously green I find myself leaning towards the idea that it is an enchanted colour. Duality fits the green profile. And again, death hovers nearby — this time in the undergrowth. The Brothers Grimm have another story about death:

> 'Be silent,' replied Death; 'have I not sent you one
> messenger after another? — did not fever come and
> seize you, shake you, and lay you prostrate? — did
> not giddiness oppress your head? — had you not
> gout in all your limbs? — did not a singing noise
> injure your ears? — had you not lumbago in your

back? — a film over your eyes? — Above all, did not
my dear half-brother, Sleep, remind you of me
every night when you lay down, as if you were
already dead!'[23]

There's a natural phenomenon called a 'green flash'
sometimes seen in clear atmosphere at the instant when
the upper rim of the sun finally disappears below the
horizon. Green is the last apparent colour from the sun
because the more greatly refracted blue is dispersed. One
of the places to see it (if you're lucky) is the Florida Keys.
Regular sunset watchers crowd into Mallory Square at
Key West most afternoons. The mayor of the town
claims to have seen it, but only once. Like the tribesmen
sitting at the arrow poison cauldron, these believers
bring a pragmatic expectancy, like a packed lunch, to the
show.

Chapter Ten

Romancing death

KILLING WITH POISON is often cast as a passionate, even romantic, act. Popular writers (as opposed to, say, coroners) know there is an audience for the murky details of a good murder scene. Feature articles, like one I found recently, hand their sophisticated post-modern audience a serving of good old-fashioned poison on a plate, and they love it. Titled 'Doctor Death', this one tells the story of Dr Shipman, the white-bearded English family practitioner accused of poisoning a lot of elderly patients with lethal injections of morphine.[24]

'In this quiet safe place the alleged crimes of Dr Shipman are all the more shocking, all the more sinister. Evil was here among them and sitting at their dinner tables. He was treating the grieving husbands while he was allegedly killing the wives. How could this have happened? How could a psychopath be so nice?' The quiet place referred to is the town of Hyde, Manchester. The town, the author writes, has a 'shivering river', 'naked black winter trees in the cemetery', 'bloated mist up on the moors' , 'muted afternoon gloom', 'grim Victorian architecture', and 'red brick terrace houses with secretive, disapproving lace curtains'. Above the story is a photo of the doctor himself, glaring at the camera, with the caption, 'Doctor Death: For 20 years he was the epitome of the old-fashioned caring family physician. Did he then begin killing his patients?' The evidence seems to stack up that way. We're told that he signed all the death certificates himself, never allowed postmortems and apparently profited from at least one of the deaths. His motive (they have a criminologist and profiler on the job) might be anything from misogyny to greed. The author interviews as many locals as will talk on the record. Her best find is the Irish priest. He dishes up Myra Hindley and Ian Bradley burying tortured children on the moors, inspiring this sentence: 'The moors, those mottled hills, rear up spectrally above the town, full of dark crevices and ominous tors.'

The feature is a pretty standard journalistic treatment for an interesting contemporary mystery. The writer's job is to define the characters, set up the divisions between them and paint in the scenery. On the one hand she has a spooky doctor who won't talk; on the other any number of willing busybodies — a chatty priest, a chap living opposite the cemetery who's been watching the exhumations and drawing his own conclusions, a crime correspondent with psychological theories, and pictures of benign-looking old ladies, labelled 'Victims One and Two'. To fill in the gaps, the author has taken a tour of the town, noting detail. She gives us the chill of the late afternoon streets coming up through her shoes, and the sound of a 'sorrowful' trumpet from the local pub. And who can argue with this sort of narrative treatment? It's the composition of details (like a pre-Raphaelite canvas) that bring the story to life; we crave them as George Orwell craved them, we want the exactness of trumpets and gap-toothed old timers holding up the bar, we want to be frightened by Doctor Death from the comfort of our lounge-chairs and for some remnant of that thrill to linger after we put the magazine down and move on to the next thing.

This is what romantic writing does: it manipulates the detail. The smallest thing can be turned to its advantage, so a reader can pick up on a quote, like ' "His [Dr Shipman's] mother had cancer when he was young," says Father Denis, searching for a reason. "She died a slow and painful death," ' and spin the possibilities out. The trouble with detail, however, is you have to get it right.

The next edition of the magazine carried this comment in its Letters column:

> I read your article ('Doctor Death, Feb 27-28')
> hoping to glean a few facts. There were precious
> little [sic], only a multitude of flowery verbs and
> adjectives. One can only assume that the author has
> some flair for fiction. Certainly not reporting. The
> fact is Hyde, Manchester, is in the heart of
> Lancashire and not within cooee of the Yorkshire
> Moors.

Oops, you might say. But does it really matter? In the greater scheme of things, isn't the journalist just giving the public what it wants — a break from the slash-and-burn killing sprees, a good yarn based on research and resting on a bed of 'flowery verbs and adjectives'?

The letter writer can't accept the whole if one part of

the whole is wrong — and there are many like him or her.
The romantic vision stands or falls on its attention to
detail. The inclusions have to be credible. 'God,' said
Robert Hughes of the pre-Raphaelite painters (but it
applies across the board to all romanticised works), 'was
in the details',[25] and God help the artist who got them
wrong.

⊳−⊶−○−⊷−⊲

As I've become more familiar with the tale of the two
murdered boys as told by my great-aunt, the suspicion of
a romantic spin on the reporting has sent me back to the
details looking for inconsistencies. Sometimes the
disparity is not in factual accuracy but in *emphasis*. If a
detail becomes too prominent it distracts from the main
game. Distractions are a deliberate device in romantic
novels: the story unfolds, cough by cough, across
hundreds of pages of plot and subplot that seem (superficially at least) to be hurrying towards a climax but are
really prolonging the moment with nimble foreplay.
Diversions also occur on canvas, not always intentionally.
Romantic painters can get carried away with fascinating
asides, as Delacroix did in his famous massacre scene,
The Death of Sardanapalus. The artist, we're told,
intended a dark death scene (the ancient king, half reclin-
ing on a bed that will soon be his funeral pyre, orders his
most precious possessions to be destroyed in his sight
before the enemy army enters the city). The king is the
centre of meaning in the painting, but as the artist
crowds in more of the story — the terrified horse in its
gold livery overpowered by the black slave, the dead and
dying slave girls, the shot and decapitated elephant —

the central idea gets lost. The details overtake the story. Irresistibly, the peaches-and-cream slave girls catch the eye and hold its gaze so the scene collapses onto its opposite face, away from death to the delights of the flesh.

The central image of our family story is the murdered children, Thomas and Patrick Macbeth. The picture of two poisoned boys ought to command all my attention but I'm drawn away by the commotion of the surrounding detail. Take the little tableau of the first meeting between William, Ellen and Rose, off to the side and painted (by Rose) in neutral browns. Those browns actually betray a sort of sexy velvet texture. Despite Rose's nudge to me to look the other way, I've had a glimpse of two young sisters, blazing with youth, bending their heads to the touch of a faith healer. It's too tantalising to pass over.

The placement of William's hand is crucial. It gives me the words that Rose couldn't say about that night: *he picked Ellen, not me.* Admitting to being overlooked is like saying 'I wasn't good enough or pretty enough' on the day, and Rose wasn't capable of that sort of concession.

To find out more I spoke with Betty, Rose's youngest sister, by telephone. Betty told me in the voice of someone who is tired of a topic, 'Rose was illegitimate, I think that had a lot of bearing on things.' I tried to picture Betty sitting in her house in Orange, put on the spot by my questions. Twenty-five years younger than Rose, Betty was an infant when the murders took place, but over time she's caught up with history and knows all the names and dates of who died, who married whom, and when the children were born. And, during the last decade of Rose's life, Betty got close to her hard-boiled,

unlucky-in-love sister. With a certain amount of disinclination, she opened the sealed package of Rose's parent-
age for me.

Rose was two years old and fatherless when a prospec-
tor called Sid began seeing her mother. Sid had worked
his way up from the south, following a vein of quartz
which was rumoured to end in gold. It didn't, but Sid
stayed anyway, not having any particular plan for his life
and finding the yellow flats by the river to his liking. He
married Rose's mother in 1903. A loner by nature, he
managed a show of placid acceptance for the kid who
came with the marriage. (Rose told Betty her real father
was a red-headed Irishman who passed through the goldfields in 1900.) Sid built a house and got a paying job.
After a 'decent' interval, Ellen was born. Whatever
ownership claims Rose had staked on Sid, she lost out
badly to baby Ellen. Sid was excited by his first experience of biological fatherhood.

Isolation and deprivation, the twin hallmarks of early
post-Federation life, pinned them to a small holding with
few prospects. As his house filled with daughters, Sid
grew distant. Rose was co-opted by her mother to help
with the younger children. She stuck close to her
mother's side, filling the companionship void left by
Sid's long absences, racing to her peak early like a forced
flower.

At sixteen Rose took a job as a domestic. She slaved for
an English family who thought she was quaint, like the
bandicoots that robbed their root vegetables. Constant
exhaustion kept her faithful to the promise she had made
her mother about boys; there would be no red-headed
Irishmen dawdling on her path. When she came home on
days off she brought saved-up stories about the English

family along with little bits of pocket money for the family pot. It gave her a small fragile edge on Ellen's developing looks.

The year Ellen turned eighteen, Rose was a legal adult, no longer in service. If Ellen left the visible limits of home territory, Rose chaperoned.

It was Rose who read the notice about the faith healer in 1922, Rose who arranged Ellen's first trip on the train and Rose who was passed over in the laying on of hands. Having saved herself for a moment just like this, having primed Ellen for careful chastity as she herself had been primed, she suffered the ignominy of being there at the precise moment when Ellen and a complete stranger fell in love.

Ellen and the faith healer William Macbeth married six months before their first son was born. I know this from dates on the certificates. By my calculations Thomas must have been conceived very close to the night they met. Rose (and Betty) glossed over this detail, even decades after it really mattered.

Betty let go of another sealed package during our brief conversation: 'The way Rose used to talk you'd think she fancied William.' From someone as reticent as Betty, the comment has the force of testimony. Rose's desire to be noticed by William (transposed with Freudian delicacy into a description of William cupping *Ellen*'s chin) found expression later when Ellen was preoccupied with a sick baby and William needed help in the business. Betty's choice of 'fancied' over more conventional phrasing like 'was attracted to' or 'liked' hints at transgression. 'Fancied' invites other readings, maybe multiple readings, of Rose's disappointed sympathies.

In pharmaceutics we say that two or more miscible

liquid phases will eventually produce a completely homogeneous mixture by diffusion. Where the degree of miscibility is small a greater effort is needed to achieve a good liquid mix. Over various stages of her life Rose exhibited consistent non-miscible characteristics. My father has a photograph taken on the Nullarbor Plain in 1972. The interest is in the grouping of the travellers — my parents to one side, sunburnt, happy; another couple sitting close together in the shade; and my father's Aunt Rose, defiantly alone. Using a magnifying glass I can see the expression on her face. It's as hard as the landscape. In forty-degree heat she's wearing crimplene slacks and a cardigan. The other couple are Rose's daughter and son-in-law but there is no connection by gesture or expression between them and the separate woman.

Recalling the original interview with Rose nearly two decades ago, I can now see her positioning herself at the centre of the action. She inserts herself into William and Ellen's early married life. She helps them find a place to live; when William needs assistance bottling medicines she's there; when her back is turned, the children are poisoned. It's hard to prise her out.

However, my challenge to some of the details of Rose's story has arrived too late. Any dissatisfied thoughts I might have about distortions and misapplied emphasis have to be addressed to the past.

>-+-+>-0-+-+-<

Elihu Vedder, who romanticised dying in paint and charcoal, painted *The Cup of Death* at least three times between 1883 and 1911.[26] Vedder met Omar Khayyám's translator Edward Fitzgerald in the summer of 1871 and for the next twelve years Vedder conceived and executed about sixty-five full-page plate illustrations to accompany the *Rubáiyát*. He changed the order of the quatrains, grouping them thematically at points of emphasis that supported his own artistic interpretation of the Persian poet's ideas, and where their philosophies parted company (Khayyám's fatalism, Vedder's attachment to Christian promise) Vedder stuck to the human dilemma and left considerations of the afterworld unresolved. *The Death Cup* (which faces a plate of its opposite, *The Love Cup*) is Vedder's realisation of the quatrain dealing with the coming of death:

> So when at last the Angel of the drink
> Of darkness finds you by the river-brink,
> And, proffering his Cup, invites your Soul
> Forth to your Lips to quaff it — do not shrink.

In a preliminary drawing for the illustration, the Death Angel is a messenger with an onerous task. Vedder's pictorial idiom is classical but his theme is romantic. The angel is a beautiful winged male in pale Grecian robes, eyes downcast. The girl in his embrace is in a half-faint. Her posture of surrender suggests courage, even transcendence (a nod to Khayyám). The angel can't even look at the girl he's about to despatch, and though their knees touch they maintain a dainty non-erotic distance. It is an intimate relationship without power; the two of them are in this together.

Vedder, more a sentimental Victorian than a pre-Raphaelite, painted another copy of *The Cup of Death* for Miss Susan Minns of Boston, 'whose fad is to have the

greatest collection of dances of death going'. Liberated
from the context of *The Rubáiyát* the artist made significant changes to his composition. The angel of death
becomes a lover. His thigh and the girl's knee press into
each other. The visual dynamics lock the figures in an
unambiguous erotic closeness. Vedder's round-breasted,
full-hipped girl in a flowing classical, off-the-shoulder
gown has her toe on the brink of the river. A few more
steps and she'll slide out of the painting, sparing us any
confrontation with the messy mechanics of dying. With
successive re-workings of the image Vedder moved
further into the idealised territory mapped out by his
own romantic notions of death.

Romantic (and fairy tale) deaths are easily slipped
into, like a warm bath. They come early in the normal
life cycle, or in a disguised form such as dissipation,
despair or insanity (presumably survivors into middle
and old age slide out of the category). Unlike some
other models, Shakespeare's for example, a romantic
death is incompatible with ugliness or anatomical
grotesqueries. The maggots and worms belong in
another tableau. Dumas' romantic lead, Armand Duval,
gets a whiff of 'a painful odour' at the disinterment of
his *dame aux camélias*. Dumas allows his hero a quick
look at the green hollows of the skull cheeks and the
eyeless sockets then rescues him with a fainting white-
ness that turns into brain fever. After a long sleep and
the onset of spring, Armand awakes restored. He
banishes the sombre picture, makes it go away, carpets
the grave with white camellias and by the accretion of
story and reminiscence succeeds in burying what he
proposes not to see. The exhumation does not blaze at
the centre of the death story. Like Delacroix's dying

tyrant, it is lost in the congestion of superimposed gift wrapping.

In order to end up as full-bodied nineteenth-century poseurs, love and death have had to travel a long way. Of the two, death was in most need of the full make-over and public relations treatment. It had to lose its late medieval image of cadavers, skulls, cobwebs, dark stars and the inescapable finger of fate. The new death, no longer the grimmest of reapers, now carried a bouquet of camellias in one hand and opera glasses in the other. Romantic *dying* became the triumphant climax of a personal odyssey. The Great Seducer swept in, scanned the crowd looking for *you,* then zeroed in like an attentive host. Foucault looked at this nineteenth-century shift and saw 'death leaving its tragic heaven to become the lyrical core of man'.

Shakespeare's death vocabulary balanced supernatural intervention with the realities of *post mortem* decomposition. His vision is graphically inclusive: even with death lying on Juliet 'like an untimely frost', Romeo can simultaneously entertain the notions of a body that has honey breath, crimson cheeks and fair skin and a body that is home to worms silently decomposing the flesh from within. When Friar Lawrence offers Juliet a draught of 'a thing like death', there isn't a winged angel holding a death cup in sight. Love and death issues are catered to by earthly priests and fatal aspects. By line six of the play we know the lovers are 'star-crossed' and therefore doomed. At the end, the human drama tumbles off its high-wire into the safety net of heavenly game-playing and the action is brought to a satisfactory close. 'Heaven,' we're reminded, 'finds means to kill your joys.'

Poison is a useful, well-understood agent of destruction in Shakespeare's dramatic world. But to get a romantic spin on death and poison, I read Keats. Apart from the obvious pleasure his poetry gives, Keats conformed elegantly to the romantic model of dying young, and he knew his botany.

> Even bees, the little almsmen of spring-bowers,
> Know there is richest juice in poison-flowers.[27]

Keats, who once remarked at a dinner party, 'Newton has destroyed all the poetry of the rainbow, by reducing it to a prism,' was a willing exile from his own training in science. When he was twenty he passed his exams at Guy's Hospital and could have earned his living as a physician had he not already decided for poetry. He wrote to a friend that, 'during a lecture, there came a sunbeam into the room, and with it a whole troop of creatures floating in the ray; and I was off with them to Oberon and fairyland,' a sensation familiar to any victim of lecture room boredom.

Listen to him talking about a plant poison, wolfsbane, also known as aconite:

> No, no, go not to Lethe, neither twist
> Wolf's-bane, tight-rooted, for its poisonous wine.[28]

Keats' wolfsbane is the perfect romantic specimen, a simple garden plant with a deceitful root. The verb 'twist' in the first line evokes Circe at work, wringing toxic drops from her rocky island plants. If we had never seen an aconite root, we could imagine its tangled, attenuated shape, its underground pallor, its earthy sweetish smell.

The tendency of nineteenth-century artists to colour-in empty spaces with extra pieces of story and allusion found no favour with sharp-eyed medical men. They distanced themselves from poets, shrugged off romance, demanded that death, like life, have order and boundaries. Where poetry and painting extended the limits of their freedom into the picturesque, the morbid and the emotional, medicine was coming to grips with the case history. Medical narratives, and the subset that interests me, poison narratives, were exploring and mapping every inch of the death scene. Before the rise of forensic science, the whole body as opposed to the divided body was the subject of scrutiny. Medical language dealt with the prosaics of surface phenomena: temperature, colour, smell, posture, sites of injury, missing pieces. Now, the gaze that surveyed the corpse extended beyond what was visible. Medical adventurers began looking *inside* the body for clues, setting the parameters for what was normal and deviant; analysts began profiling fluids and tissues; and new language was minted to describe previously unseen organic changes. Death, the sexy thief, was being defrocked.

In the nineteenth century the evolutionary paths of both popular and medical narratives about how we die from poison — that is, why we believe what we believe today — underwent the equivalent of a transition from reptiles to birds. Medicine's leap forward was put into extensive, annotated case histories, carefully devoid of superstition, guesses, religion or philosophy and perpetuated in journals, texts and the lecture room. These narratives laid down an uncompromising, pathological picture of death at work that was far removed from winged angels.

If medicine hadn't been such a closed shop until the twilight hours of the twentieth confessional century, maybe more of its vernacular might have taken root in the public narrative about how we live and die. Though easy enough to find if you look, case histories are not intended for the untrained eye. The gaze that surveys the wreckage caused by poison is one steeled against flinching.

To see how science described wolfsbane (or aconite) and to start me off on nineteenth-century toxicology I took down Dr Alexander Blyth's standard work, *Poisons: their Effects and Detection*, published in 1884.

The top four nineteenth-century poisons were arsenic, coal gas, opium and mercury, but occasionally one comes across an interesting killer vegetable. Of aconite, Blyth wrote, 'there are few substances which have been experimented upon in such a variety of ways and upon so many classes of animals as aconitine [the pure form of the raw plant drug] in different forms'. Dr Blyth personally applied dabs of poison to blowflies. 'If a minute dot of aconitine was placed on the head of a blow-fly . . . very marked symptoms would result . . . muscular weakness, inability to fly, and to walk up perpendicular surfaces . . . a curious entanglement of the legs, and very often extrusion of the proboscis.' Death, he noted, was so gradual that it was 'difficult to know when the event occurred'.

Plunging into the case histories I came across Mary Anne McConkey who famously sprinkled aconite on her husband's lunch, killing him in three hours.[29] Mary Anne was hung in Dublin in 1841. *Regina v. McConkey* is a footnote to history that has been mined for years. Writers on aconite usually trot out Mrs McConkey. They lay her on the page like a legal precedent before a judge, forgetting (if they ever knew) that aconite also has a poisonous history dating back to the Greeks.

The events of the late summer day in 1840 when Mary Anne criminally transgressed are heartbreakingly domestic and familiar. The McConkeys had guests for the midday meal. Mary Anne, her husband Richard, Mary Anne's father and a neighbour all sat down to boiled potatoes, fried bacon and 'greens', variously reported as cabbage, spinach and beet tops. Mary Anne served her father and her husband first, then her neighbour and herself. Later, it was reported that she kept her back to her guests during the preparation of her husband's greens, and that she'd kept his greens separate from the communal pot.

Mr McConkey complained that his cabbage tasted 'sharp', but he ate it. What lies did she tell to explain the taste? — a comment about the barrow boy or the late rains, or perhaps she said nothing but *eat your greens or I'll throw them to the pigs*, as wives do?

Before the meal was over Mr McGeighan, another neighbour, made a late appearance. He was a senior man, her father's age, and she was duty-bound to feed him. Mrs McConkey's panic rises off the page at this point in the narrative; her husband instructed her to give the uneaten portion of his 'wild tasting' greens to Mr McGeighan, perhaps to prove a point. She insisted the spoiled food be thrown away but was over-ruled and had to spoon her husband's leftovers onto McGeighan's plate. The older man chided McConkey for complaining over the taste, saying the greens were 'good'.

McGeighan left for his own home straight after the meal. On the way he asked a neighbour if his face was puffed up and when told no, wouldn't believe it. The curious neighbour checked on him an hour later to find

McGeighan frothing at the mouth and nose, alternately passing out and reviving. A doctor was called late that evening. McGeighan spent five miserable weeks recovering before he could go back to work.

Neighbours and onlookers (who later became witnesses) visited the McConkey house too. Mary Anne's husband was in a terrible state, exhibiting all of McGeighan's symptoms but raised to a higher degree. McConkey writhed on his bed, repelling his wife's attentions. Mary Anne kept busy scraping up the body fluids with a spade and carrying them outside, actions later interpreted (correctly) as removing the evidence.

He wasn't seen by a medical man until the day after his death. Initially a mineral poison was suspected. McConkey's viscera were removed and sent away for examination but the chemical examination alone couldn't decide the cause of death. Cholera and diet were ruled out, as was arsenic, mercury, natural and epidemic disease. Non-medical witnesses were called to provide 'certain general circumstances intimately connected with the medical facts'. Neighbours and onlookers and relatives thread their way in and out of this story. How Mary Anne could even entertain the idea of quietly killing off her husband in the goldfish bowl of her local society as nosy as Cranford, is perhaps a measure of her desperation.

Mineral poisons being discounted, the animal poison cantharidin (Spanish Fly) was briefly considered. However cantharidin leaves a much messier footprint and, according to the record, was never purchased by the prisoner. Thus, by elimination, suspicion fell on the plant kingdom and specifically on aconite, or monkshood (or wolfsbane), which happened to grow wild in the district. The locals called it 'blue rocket'. Its blue flowers were used for decoration on market day. The principal medical witness for the prosecution, Dr Geoghegan, learned that the root of the monkshood's plant was well known to be deadly by the locals, including the deceased and the accused, and that stories were in circulation about how quickly a person could die if he ate it. Anything from two to four hours seemed to be the consensus of anecdotal reports. The source of this local knowledge was jockeys and trainers who were involved in the widespread, illegal practice of 'figging' a race horse, or making it more lively, by introducing aconite into the horse's anus.

Mary Anne McConkey laid her poison on thick. I think she was surprised by the suffering her husband went through before he died. I think she relied on stories and the advice of people she thought smarter and more worldly than herself.

In the final hours she found it impossible to watch with a cold eye as her husband stumbled through his slow-motion death dance — in prison she confessed and asked forgiveness.

It's a brackish little story made entertaining by familiar domestic detail. The kitchen dynamic touches us all. If we can't trust the hand that cooks the food and warms the plate, then the mappable territory of what we understand as the *hearth* — Mary Anne's pots of greens and potatoes boiling on the fire — turns into a foreign land. What is essentially disturbing in Mary Anne's case is the sympathetic tug felt by the modern reader, by myself, towards her and away from the husband. Surely Dr Geoghegan didn't intend us to feel any sympathy for Mary Anne. In fact, his learned discussion is correctly silent on the matter of her sensibilities. They have no place in medical discourse. Mary Anne, whose voice we long to hear, who has — after all — the most to say, is silenced. She is the accused. She has to face the full blast of science and the law in mute submission to higher powers.

The McConkey case is cited in a 1958 paper in the *British Medical Journal*. Dr Fiddes MD reports on a story of two medical students who tasted the contents of some small glass phials they were supposed to be clearing out. Each got a dose of aconitine and suffered vomiting, numbness, salivation, and feelings of dread. At one stage their pulses became undetectable, then they recovered and went on to give Dr Fiddes enough information for a paper.

Dr Fiddes transliterates the foolish actions of the students, Mr B and Mr T, into the required medical journalese: 'Mr B. experienced a sharp, slightly bitter taste, and a rapidly developing sensation of warmth and tingling in his tongue, which increased in intensity and extent until his whole mouth was affected.' Notice the echo of the adjective 'sharp' from the McConkey case. Then Dr Fiddes does an incredible thing: in his paragraph headed 'Identification' he makes the admission that other than *trying the drug himself* he can't say with any certainty that the substance is anything other than an unspecified alkaloid: 'The substance was personally sampled, perhaps slightly to excess'! — with the result that 'for some time' (*read, 'more time than was comfortable'*) the doctor's mouth and throat became numb and tight and his fingers tingled. Given that the spectrophotometer and gas chromatograph had been around for about four years before Dr Fiddes's paper (his conclusion gives a nod in technology's direction) one wonders if a bit of reckless curiosity overcame the medical man.

In fact Dr Fiddes was conducting what was called, a hundred years before his time, a 'life-test'. The state of knowledge of alkaloids back in the nineteenth century, though brisk in endeavour, had been 'deficient in delicacy', and some of its less sophisticated techniques still prevailed in otherwise modern laboratories. In order to clinch the identification, Fiddes injected half a milligram subcutaneously into a rat, noting, 'After a delay of about a minute, the animal showed signs of restless- ness and agitation, and started to run. Salivation, retching, and respiratory distress developed, followed by paralysis and terminal convulsions before the animal died three minutes after being given the injection.'

One of the most dramatic examples of the 'life-test' was that by Dr Meyer, who treated a patient with aconi- tine drops in 1880. The patient became so ill from the treatment (extreme chills, loss of sight and hearing, spasms, convulsions) his wife returned to the doctor 'and accused the medicine'. But Dr Meyer would not be criti- cised. To prove his argument he took a dose from the patient's bottle, and died of aconitine poisoning five hours later. First reported in a Berlin medical journal in the same year, this case delighted British pioneers in toxicological reportage. Here was a life-test *carried to extreme* with every dose and detail faithfully recorded as if it had been an intentional experiment in the limits of human tolerance to poison.

><><><><><

Medically, death is recognised in two phases. The first is somatic death. The personality disappears, the vital processes halt, the person we loved and knew becomes a corpse. The second phase is molecular death. Progressively, in a certain order (intestine followed by stomach, liver, heart, blood circulation, heart muscle, air passages, lungs, brain, spinal cord, kidneys, bladder, testes, voluntary muscle, uterus, prostate), the tissues, organs, glands, tunnels, vessels and the great watery vat of fluids that sustain life, all begin to disintegrate.

Medical writers have laid down a cold white corpse. In their efforts *not* to romanticise, to elevate fact over supposition, to separate the deed from the doer and to silence all other voices but their own, they've extruded a parallel layer of expression, separate and discontinuous from the romantic layer.

Perhaps this is why we have so many metaphors for dying.

Last year my friend Jenny had her pulled-muscle pain investigated properly. It was lung cancer. Within three months she'd developed the cachectic look of someone close to the end. Her recounting of horrible clinical procedures was part text-book, part interested observer, part participant in a sporting event that wasn't quite over yet. In none of this talking to me was the word 'death' ever spoken.

The last weeks turned sour. She retired to her upstairs bedroom with an oxygen bottle and a live-in nurse. Morphine was her best friend. She laid bets on the outcome of the next blood transfusion. At what point, I wondered, does one of us mention the D-word?

In the last week her house was full of relatives sitting vigil. 'I think we're reaching the climax,' she said and died four days later while the nurse was changing her nightgown.

After it was over I asked questions. I learned that her

father had died of lung cancer twelve months before and was death-denying to the end. I learned that she'd cried with her best friend, howling at the injustice of getting only fifty-one years and not seeing a grandchild. From her priest I learned she'd gone looking for her lost God and found him.

It came naturally to Jenny to choose sporting-metaphor language for her death narrative. She was tall and lean from years of playing tennis and squash. She had a ready-made vocabulary for winning and losing, for trying harder, getting fitter, going one better and picking herself up after a defeat. She had a competitive edge, she liked to win. The metaphor held. And it worked, for her. She died in her pretty bedroom with her carefully positioned pretty things in view. She died on a sunny day, going down for the count because she had to, because her opponent was better on the day, because she didn't have enough fight left.

You could call her dying a sort of poisoning. Twenty cigarettes a day for twenty years adds up to a lot of nicotine and tar, each dose in itself not lethal, but cumulative through repetition, damage on top of damage, and the passage of time. That much nicotine taken in one go would have sent her to heaven in seconds. I went to an old toxicology book and looked up poisoning by pure nicotine. It's rare. You can't get hold of it, unless like Count Bocarmé in the 1850s you study chemistry and learn how to extract it from fresh tobacco leaves. Even if you get your few drops of oil, its hideous smell and taste make it too noticeable to hide. The Count had to force it down his wife's throat. She died of respiratory paralysis in five minutes.

Nearly every fatal case of nicotine poisoning is accidental; but criminal cases are not unknown. From Dr Blyth's casebook we learn of a ten-week-old child who was killed when her father pushed tobacco into her mouth. A drunken sailor who swallowed half an ounce of tobacco in one gulp and endured a giddy seven-hour fall into death. A three-year-old child who blew soap bubbles through an old tobacco pipe and died a few days later. And two German brothers who had a smoking competition, one pipe after another until they got to seventeen when the first one died, followed soon after by the other after eighteen.

Death by cancer allows us to see the dark angel at his work. It can take years to die from the inside out. While this dying is going on, parts of our bodies that used to sit somewhere quietly doing their jobs are hauled out and examined. The lobe of a lung might be cut off and thrown away. A lump of breast might be sent to the lab and put to the test. By choosing to have chemotherapy (over no therapy) Jenny opted for a societal norm, that is, she chose remedial action to bring herself back to her normal self. Part of this package included the idea that remedial action might fail (it did), which then entitles the user to proceed to the socially sanctioned state of being beyond help, followed by death 'as a social accomplishment'.

⊷⊶⊙⊷⊶

Before we leave the world of the medical case history, I want to re-tell a twentieth-century story of attempted romance gone fatally wrong. A male chemist's assistant in a British firm of chemists had his eye on a young female clerk working in the same store. Perhaps his approaches were rejected, or perhaps he was too shy to even attempt an approach, but whatever the reason, he chose to seduce her with an aphrodisiac. I felt a strange (unhappy) kinship with this unnamed young man, having once worked as an apprentice inside a white coat. As an unqualified assistant he would have spent his days in mundane tasks — mixing, measuring, taking pains to get his quantities correct, submitting all his work for the approval of a busy senior qualified chemist. It's natural for the mind to wander in this claustrophobic world that's a bit like a wizard's chamber. There are remedies for everything to hand (except the thing you might desperately want, like escape or autonomy).

The assistant read up on aphrodisiacs and discovered Spanish Fly, a drug with an anecdotal reputation for seizing its recipient with sudden overwhelming attacks of desire. Mr X, as he was called, asked his senior chemist for a quantity of the substance, using its chemical name cantharidin, saying he wanted it for someone else, a neighbour who bred rabbits. Sensibly the senior chemist refused to hand any over, but during his lunch break the young man (who now knew where the drug was kept) stole a quantity and hid it in an envelope.

He bought eight pieces of coconut ice and put a small amount of cantharidin into two squares. One was intended for the girl, the other, one presumes, for himself. At afternoon tea he offered the treats to the clerks. The girl ate her adulterated piece, and in what was described as an 'incredible piece of carelessness' a senior woman ate the other spiked piece.

About ten minutes later the two women felt ill. The older woman asked for a taxi to take her home and during the ride she collapsed. Her doctor was called and he arranged for her to go to the hospital. It was over seven hours between eating the coconut ice and reaching the hospital and the woman was by this stage bringing up pure blood. The younger girl, the object of Mr X's attention, was in an even worse state than her unfortunate colleague.

Both women died from blood loss.

The post-mortems revealed extensive internal bleeding. The entire mucosal surface from the tip of the tongue to the stomach had been stripped away. Lower down, the kidneys, bladder and ovaries were blistered and bleeding. The heart and lungs were similarly affected. The lower part of the oesophagus appeared to have been shredded, and even the gentlest pressure caused strips of mucosa to come away. The apprentice was hospitalised with blisters on his face from touching the crystals. (The old 'blister test for cantharidin poisoning' — before chemical assays came into being — was to obtain stomach fluid from the corpse and apply it to healthy skin). He was gaoled for five years.

The young man stupidly and disastrously mistook the poison cup for the love cup. The cup of love is a mythical leg-opener that goes back in time to witches' philtres and forward to Viagra for women. Blyth reports a nineteenth-century Viennese experiment with cantharidin. A physician took thirty freshly killed blister beetles (Spanish Flies) and soaked them in alcohol for a fortnight to make a tincture. He drank ten drops in water and soon felt a pleasant warmth sweep through his body. He then enjoyed three hours of 'sexual excitement' but this was soon followed by pain, diarrhoea, straining and frequent urination (with pain). The doctor repeated his experiment, this time using pure cantharidin with a much increased dose — and dire results. It took two weeks to recover from a flayed tongue, swollen glands, bloody urine, intense abdominal pain, a racing pulse. Blyth concludes that the 'popular idea of exciting sexual passion holds good only as to the entire cantharides, and not with cantharidin'. Nasty little stories about Spanish Fly pop up all through the literature of poisoning, usually in the context of an attempt to sexually arouse an unwilling or apathetic 'other'. Occasionally there is a strange deviation from the pattern, like the fisherman who mixed cantharidin in water to soak his bait in. While shaking the bottle he contaminated his finger. Later he pricked the same finger and sucked it. Two days later he died after an ordeal. The Spanish Fly was intended to make the bait 'sexy' for the fish.

▷┄◈┄○┄◈┄◁

Tantalising additives bring me back full circle to my grandfather. When he made a tonic *For gentlemen only,* he enlivened the original formula with a pinch of strychnine. The recipes for patent medicines are deliberately obscure but the principles are universally known. Quackery by definition didn't bother itself with expensive or hard-to-come-by ingredients, it sold an idea: 'Manhood Restored', 'Your ability to fulfil *man's most sacred obligation* regained' — these are the mottoes of popular wish fulfilment.

Strychnine is part of William Macbeth's story, there is ample evidence for its existence, but as in all narrative edifices, placement is everything. Strychnine is a *detail* that sent my research down a jungle track in search of an exotic nut with a bitter seed. It now puts me in mind of the moors: the moors in the *Doctor Death* article:

realistically they couldn't exist, stylistically it mattered to have them.

The general reading public come by their poison most often through journalism and fiction, rarely from coroner's reports or medical journals. The two streams that came down from the nineteenth century mingle only minutes before they reach the sea. The *British Medical Journal* once remarked in a book review that the public

> seem to cherish an odd affection for those old malefactors who prey upon them. Most modern cases of poisoning are quite devoid of romantic features, and there is little of that subtlety which we associate with the efforts of Medea, of Locusta (Nero's adviser and collaborator), of Toffania or other members of the Roman and Venetian schools of toxicology. Possibly, of course, time operates and we see those ancient murderers through rose-tinted spectacles. The author [*of the reviewed book*], no doubt wisely, does not attempt to dramatize, but makes the point that poisoners are in a general a sordid and miserable lot.

I give the final word against romantic notions of dying to the late Philip Hodgins, an Australian poet who gave the matter much consideration. In 'Ten Things About It' ('it' being his impending death from leukemia), from *Blood and Bone* 1986, his point number three is 'The nineteenth century doesn't come into it.'

Chapter Eleven

The magician's clothes

WHAT SORT OF CRIMINAL was William Macbeth? According to his wife's family he was the worst kind, a man who killed two children and walked away scot-free. In retaliation the family erased his voice from the oral record that supports and shapes the murders, without, it seems, ever exercising the greatest reprisal of all, calling in the law. The official records which ought to scream with outrage at his crimes are unalterably silent.

In fact, the question of William's guilt has only ever been addressed from the perspective of his detractors, in private. If there'd been a court transcript as in other poison trials, some of the distortion of perspective might have been corrected. We might have heard William's voice honed to its clearest expression as he fought for his life and reputation in the dock, mounting his defence on the ramparts of reasoned argument. There might have been days of expert testimony on the potency of strychnine and a chemist's duty of care. The truth, if it emerged, could have been a simple statement of neglect, a moment of inattention; or the ultimate finale to speculation, a frank confession in the face of overwhelming evidence.

If William was a particular *type* of poisoner, you could say he was like Dr Crippen. Both were medical privateers, both specialised in patent medicines and both blurred the distinctions between medicine and pharmacy (and coincidentally, dentistry). In the crucial matter of consequences, however, their destinies were markedly dissimilar. Hawley Harvey Crippen was hanged for his sins in 1910.

I found Crippen a long time before I found William Macbeth.[30] There's a well-worn path to Crippen's door. The grooves are so deep you can plainly see where others have tramped through the written records in the hope of catching a glimpse of what went on at 39 Hilldrop Crescent on the night of 1 February 1910. The only witnesses were Crippen himself, who always denied that there had been a murder at all, and his wife who was never seen alive again. Beating my own path to the Crippen monolith I felt the passionate condemnation he received from the prosecution, a hangover no doubt from my early brushes with William's accusers, and was conscious also that there might be something useful to learn from the detail surrounding the central drama.

Hawley Harvey Crippen (Peter to his friends) got his medical degree at the Homeopathic Hospital in Cleveland, Ohio. He had never conducted a post-mortem in his life and only ever attended a theoretical course in surgery, aspects of his training which were picked over in considering the medico-legal implications of the *professionally* carved-up female body in his cellar. His area of special interest was the eye, ear, nose and throat but he was most often employed in making and selling patent medicines. (He also had a small mail-order business in homeopathic remedies.) Any dexterity he lacked with the scalpel was made up for with his meticulous dispensary skills. Questioned about a purchase of hyoscine, the poison found in the remains of the woman in the cellar, Crippen describes how he prepared the drug for mail-order clients:

> 'I dissolved hyoscine crystals in alcohol, and then
> I dissolved 5 grains in an ounce of water, that is, in
> 480 drops, giving to one drop 5/480ths of a grain.
> I used four drops of this, which would equal
> 20/480ths or 0.041 of a grain, in conjunction
> with another mixture consisting of gelsemium,
> asafoetida, and some other homeopathic
> preparation. This, with a drachm of the other
> mixture, I used for medicating 300 small discs.'

Asked, 'Did you at any time ever administer hyoscine to your wife?' Crippen replied, 'Never.'

Hyoscine belongs to the deadly nightshade or toxic potato family. In 1910 the chemical extract of the plant was not in general use in England and there seems to have been an air of mystery about this 'new' poison (the chemist had to make a special order to meet Dr Crippen's request). Yet the whole plant, henbane or *Hyoscyamus niger*, had been known since ancient Greek times for its therapeutic pain-relieving and soporific properties. And had always had a dubious following for its non-therapeutic uses: the induction of visions and altered states.

Crippen was a great fan of hyoscine. In America he'd seen it used in insane asylums for quietening patients with mania. In homeopathic doses he found it to be a creditable nerve remedy, while in medicinal doses it eased the coughing of asthma. He also used it in his eye clinic, to enlarge and paralyse the pupil for examination.

The prosecution contended that Crippen gave his wife a fatal dose of hyoscine late one evening after their dinner guests had left, then dismembered her body, disposed of some parts (the head, limbs and skeleton were never found) and buried the remains in the cellar. The next morning he left for work at exactly the same time in a normal, unremarkable manner except for one unusual excursion: he called on his dinner guests to inform them that his wife had suddenly left for America.

Digging into his past and finding a first wife (deceased, natural causes) and child (now living with grandparents) in America, the prosecuting counsel wove the threads of his second marriage to nineteen-year-old Cora and his, much later, affair with Ethel le Neve (who never quite got to be the third Mrs Crippen) into a sort of repeating pattern of violent and passionate attachments to strong women. Crippen, it was contended, killed Cora because he fell out of love with her and was desperate to bed Ethel, who was holding on to her virginity until she saw a wedding ring. It must have been a hard stretch of the jury's imagination to match this constructed image of a serial Lothario to the forty-eight-year-old doctor in the dock. Crippen as the monster at bay was a sad disappointment to the crowds who'd queued for hours for a glimpse. A myopic little man, five feet four inches tall, he stood meekly in front of his accusers blinking like an upset goldfish. Alcohol soured his digestion, tobacco made him cough, and his 'eccentric taste in neckties and dress' was explained by the fact that Cora (who had aspired to a music hall career) always bought his clothes and dressed him.

Ultimately it was hard forensic evidence that sent Crippen to the gallows. Cora's hairclip, pyjama top, hair strands, and an undecomposed piece of her skin bearing an identifiable scar, convinced the jury that the woman in the cellar was Mrs Crippen, and in twenty-seven minutes they returned a guilty verdict.

Crippen achieved legendary status (there are webpages, a wax effigy at Madame Tussaud's, trial exhibits at New Scotland Yard) by *getting caught* in the full glare of public attention after some very clumsy post-murder behaviour and a chase across the Atlantic Ocean. In one of the earliest examples of breaking news, his whereabouts were telegraphed by the latest technology, a Marconi wireless set, to police, who boarded a second ship and outsailed him to Quebec.

Murdering doctors attract nervous attention from the public. Even if we know doctors to be fallible humans like ourselves, we nevertheless relate — via these flesh-and-blood practitioners — to an archetypal Divine Healer whose interest in our wellbeing is limitless and benign. When a representative of the Divine Doctor sins against one of us we go into a collective shiver. One of the coldest fish in this sinner category (significantly mentioned by George Orwell as one of a list of seven murderers 'whose reputations have withstood the test of time'), was Dr William Palmer — a serial poisoner of relatives and friends alike, who was hanged in 1856 after a celebrated trial in which defence counsel spoke for eight hours.[31] Sir James Stephen, writing about the case, said,

> His career supplied one of the proofs of a fact which many kind-hearted people seem to doubt; namely, that such a thing as atrocious wickedness is

consistent with good education, perfect sanity, and everything, in a word, which deprives men of all excuse for a crime. Palmer was respectably brought up; apart from his extravagance and vice, he might have lived comfortably enough. He was a model of physical health and strength, and was courageous, determined and energetic. No one ever suggested that there was even a disposition towards madness in him; yet he was as cruel, as treacherous, as greedy of money and pleasure, as brutally hard-hearted and sensual a wretch as it is possible to imagine. If he had been the lowest and most ignorant ruffian that ever sprang from a long line of criminal ancestors, he could not have been worse than he was.

Poison is an ever-present temptation to anyone with a destructive inclination who also happens to work *inside* society's barriers to its possession. Doctors, pharmacists, nurses, laboratory assistants, photographers, miners (and many others) have a legal right to deal in toxic chemicals and plants. The means are simply *there* for the taking. All except a small handful of doctors regularly work in the terrible brightness of expediency without giving it a second's thought. It's the small minority who load the syringe with an overdose knowing that a sudden (or slow) change in a patient's health is for them alone to interpret and act on, and death is for them to certify.

Dr Crippen murdered, they say, expediently. Cora's death wasn't symbolic, ritualistic or sociopathic. A writer of the time called it a *crime passionnel,* implying something unBritish, something a hotblooded foreigner would do. I see another layer under this opportunism, something more than just insider trading. Homeopathy, the bedrock of Crippen's training, is predicated on two laws, the law of similars, *similia similibus curantur* (let like be cured by like), and the law of infinitesimals, where the pure drug is serially diluted to its smallest yet paradoxically most potent dose. In Crippen's American experience, homeopathic doses of hyoscine calmed mania and the association of pure drug with pure effect became a motif in his thinking. Under his own roof, seventeen years into a loveless marriage, Cora, he contended, had become a bullying madwoman, spending his money, neglecting the house, denying him sex yet taking lovers. From records we know that Crippen regularly bought morphine, cocaine and mercury — all normal adjuncts to his work in ophthalmology — at his local chemist, and could have purchased any quantities of these poisons without challenge simply because he was a doctor whose identity had been established and accepted, but that on 17 January he ordered five grains of hysocine, with, the prosecution contended, deadly intent. 'How dangerous it is for an ill-natured woman to be married to a physician,' wrote Voltaire.[32] It would make homeopathic sense to Crippen to silence Cora, the lunatic in his house, with hyoscine, his version of Voltaire's 'affectionate remedy'.

▷─┤◆〉─○─〈◆├─◁

The Crippen case was a bit over ten years old in 1921 and had been widely covered in the Australian press when Dr William Macbeth launched his faith healing career. A lot is known of his life from 1921 onwards, but discovering the early years has been a hampered meditative process: hampered because the usual channels of birth, death and marriage records don't work when names have been changed and living memories have failed, and meditative in the sense of years spent brooding over notes, trying to see William's shape in the unprovable belief that I will know the outline when it appears.

My vocational decision to enter pharmacy had nothing to do with recondite links to my grandfather. I graduated before I knew what his profession had been. Only in retrospect does it seem apparent that by going along this path I was turning myself into, in a sense, the ideal 'reader' of William's life. What's left of William's possessions — his suitcases, collar boxes, calling cards, newspaper clippings — have all found their way to me and the reasons for this are as mysterious to me as to my family. Why should I be the one to store the relics and, now, to spend years trying to reassemble the man my relatives hate?

Once, on a whim, I asked a psychometric reader to tell me what she could divine from William's monogrammed leather handkerchief-box. It was silly and non-scientific, and felt like exactly the right course to follow on that particular day. As is the case in these transactions, I paid my money and sealed my lips. After holding the box for a short time the reader told me a story about a man who loved and lost a woman. 'He died loving her, you know,' she said, opening her eyes and studying my face. 'And, he has something to tell you.' I tried to look non-committal. 'Look for the answer in the mixing.'

>─┼─◆─◆─O─◆─┼─◄

In his photographs William looks like a man comfortably engaged with the present. People who knew him used adjectives like 'calm', 'warm', and 'caring'. I once spoke to a man in Wellington who said Macbeth fixed his 'rheumatics' in half an hour. He, like Rose, also told me about the top hat and silver-handled cane. My father (William's youngest son) remembers a suit with a silk lining and the smell of tobacco and scent. Attached to every first-hand recollection I've managed to hear is the rider about his clothes, as if the two elements, the healing (which includes its opposite, harming) and his apparel, are connected and significant.

The link has its source, I'm sure, in mythical imagery: the robes of the Divine Healer and his other manifestations, shaman, nun, priest, witch-doctor. William's dandyism exploited the theatrical impact of the shaman; he was a showman from the emerging connoisseur class of the 1920s, dressed, as it were, to kill. I thought, holding up the photographs, there was some shift in thinking necessary, something not to be dismissed in the matter of the magician's clothes.

⊱━◈━○━◈━┉

When I looked again through my notes from the 1980 interview with my great-aunt there seemed to be nothing left to squeeze from those pages. I flipped through the well-skimmed sentences from which, like a geologist, I'd taken depth soundings, rock samples, acidity levels and from which I'd chipped every atom of meaning. Then a single word leapt out at me. Of course! I said to myself, why hadn't I seen it before? We betray ourselves in our vocabulary and most often with our nouns, the naming words we all learn and store away to describe the world we know. I re-read the passage where Rose described William making his tonics. After categorising his technique as domestic (the bottling-quinces reference) and his insistences as pedantic (bottle washing, working from left to right), she turned to his vocabulary. 'He made the tonic in what he called *duplex*; that was his training, he said.'

Today the word 'duplex' is most commonly heard in relation to housing, but to someone with a chemist's ear it means a double strength solution, made in advance of requirements then watered down to normal strength at the time of dispensing. The specificity of this usage and the dates put William firmly in the engine room of a working dispensary possibly in Sydney, and definitely at the time of the Great War.

If at some stage William had been indentured as a pharmacy apprentice I might find a small thread dangling somewhere in the archives, or hear a murmur on the line that connects all members of the profession to each other, no matter how hard they try to hide.

In 1900 there were only 182 chemists in the Sydney area, by 1916 there were 1200 in the state of New South Wales, and behind every chemist there were one or two apprentices toiling in subterranean holes or poorly lit back rooms making stock quantities of decoctions, tinctures and ointment bases, being paid very little and sometimes nothing at all. William had always been deliberately cryptic about his training. During the years of his marriage to Ellen, he'd said very little about his early life, and nothing at all truthful about his ancestral ties. He allowed those who wanted pedigrees to take him for a Scotsman, educated in England and America and recently landed on Australian shores.

It's not strange that Rose never referred to William's background in our talk, or that my father's relatives

spoke of William as though he'd arrived on the planet
fully formed, aged twenty-one, in top hat and tails. He
might as well have stepped off a boat from America —
which was as far away as most Australians of the day
could imagine — and, in stepping into our insular and
isolated culture, have suffered the same fate as any
immigrant: the denial that what preceded entry to this
country has any continuity of relevance. America was
the perfect smokescreen. Who among the farmers of
country New South Wales or the citizens of Sydney's
new suburbs knew or cared about a distant colony that
was once England's? Pennsylvania could have been Tran-
sylvania.

Toiling through the pharmacy archives I eventually
gave up on *William Macbeth* and looked instead at
names with the initials WM.

Amateur detective work and hunches might be the last
resort of the failed researcher but in this case, with the
help of document searches and talks with retired
chemists, I found a candidate. My family doesn't want his
real name disclosed. My father grew up with an embar-
rassment of surnames and an identity muddle that was
never really sorted out. Even if my impulse is to rip the
final veil off secrecy, appending another name seems
unnecessarily burdensome. It might be a mistaken
strategy altogether. After all, I argued, didn't Perceval get
all the way to the Grail and ask the wrong question?

In writing about my grandfather I'm mindful of two
things: whatever I offer back to the family will never fully
mitigate their view of the past, and whatever historical
accuracies I find out about this man won't affirm or
negate the prevailing narrative truth — though it might
just soften the edges. Crippen, Palmer, Maybrick,

Seddon, all the luminaries stretching back to Nero's favourite poisoner, Locusta, entered into a private bargain with the darker forces. The right question for me to ask might be: what intermediary brought William to the same bargaining table?

The breakthrough I needed came from an advertisement I placed in a national magazine. I received a letter from a man who believed his grandmother to have been William's sister. And so she was. (I received many, many letters, some excitedly detailing — with trees and maps and footnotes — their connections to this and that line of descent. One woman wrote that although the names and dates didn't match my request, and no one of her blood was related to anyone of my blood, I might be interested to see her charts.)

William's sister was Jane. (She and I coincidentally shared the day and month of our births.) She died in 1972 and left boxes of papers, diaries and memorabilia. My newly found relative gave me permission to read and take notes, without himself taking any interest in proceedings. The spoken stipulation was that I was to remove nothing. I chose not to tell him about William's crimes even though in Australia, where many of us trace our ancestry back to convicts, having a criminal in the background is barely remarkable. My father and sister, whom I'd brought along to give the day a sense of occasion, both tried to warm to this man and his family but we were all clearly uncomfortable in the artificial atmosphere. On this first visit my support team hung back like witnesses to a grave robbery.

Born at Coogee (a beach suburb near Sydney's famous Bondi) in 1899 to first generation *Irish* immigrant parents, William was the youngest of six siblings, his mother's 'darling boy'. Jane, the eldest, was born fifteen years before William.

It appears that he could have settled for a life of inherited comfort if he'd conformed, but from an early age William (and Jane) cultivated 'otherness'. Jane kept house for her businessman husband and told fortunes on Wednesdays and Saturdays, by appointment. She kept newspaper cuttings and announcements relating to William, and amongst these I found William's horoscope, on which she'd written 'Neptunian in all things'.

William's mother might have been the one to set her son's feet on the path from darling boy to poisoner when she persuaded Mr Smith, the family chemist, to take William on as a delivery boy at the pharmacy. The chemist was an upright, slightly pompous middle-aged man, still known as *young Mr Smith* to the clientele, who had once patronised his father. He was generally thought to be an inferior version of his father, old Mr Smith. Certainly he was more venal and more often than not in an irritated mood in the presence of the public. Two years later, in a climate of acceleration thrown up by impending war, William's mother asked Mr Smith to indenture the boy as an apprentice chemist.

Mr Smith, I learned, despised the general public in an indiscriminate way. He loathed the rich as much as the poor, the sick as much as the well. He loathed salesmen, tinkers, inspectors, couriers, the postman and Dr Graham, whose surgery two doors up the road provided the bulk of his prescription trade. He ignored his staff and never answered a direct question, and at each pay day he enacted a scene of pained surprise that any of them would want money from him. His business survived only because of Mr Harvey, the front-of-shop manager (most often mistaken for the chemist), and Mrs Abela, who knew everything about skin. It was from Mr Smith that William learned careful if pedantic dispensing habits. William suffered in this bleak misanthropic gloom until the last days of 1916 when Mr Smith, who'd been heard to say he could do more mischief with a pound of arsenic than a Lewis gun, suddenly enlisted.

Certain moments have a dreamlike numinous quality, even in the green still-unfolding days of a young man's life. Old Mr Smith, brought out of retirement to stand in for his son, introduced William to therapeutic touch at their first meeting. He put the palm of his hand to the scowling boy's forehead, said nothing, looked at his watch and moved on to greet the rest of the staff. William experienced a radiating weakness that caused him to sag against the bench.

In later years William Macbeth, conjurer, healer, chameleon, and preacher, retained the ability to revisit that first meeting with the old man, to re-evoke that fainting feeling followed by intense relief, a total somatic response (so he told his sister) that could be approached but never fully achieved even in carnal experience.

Almost eighteen years old, William had reached his full adult height. In his day suit, his hair shining with cream, his complexion a testimony to Mrs Abela's cleansing and toning regime, he had begun to generate the aura of higher destiny that he wore as a man until his final illness. One day, on a message to Dr Graham's, he was mistaken for the replacement doctor due in from Sydney that morning. The sensation pleased him. It fed into and augmented old Mr Smith's comment, 'Here's our young lion, head and shoulders above the rest,' establishing a consciousness of self bound to and prescribed by clothes, bearing and demeanour, that was (to him) not inferior to, but other than, a soldier's uniform. He had found a way to be a civilian hero.

During the eighteen months of old Mr Smith's tenure in the pharmacy, William underwent a professional realignment, unintended, unsought, yet far-reaching. Old Mr Smith's opinion of the curative effects of drugs was ambiguous. He had total faith in adrenaline, opium, cascara, digoxin and iodine. Wheezing chests, pain, stopped bowels, slow hearts and infections of all sorts had his total support. The rest, and this was a large amorphous area of human ailment, he viewed with a compassionate eye and treated with therapeutic touch. From all I've gathered about him, it seems the anarchic message of the old man's teaching seeded into the split-off part of William where it flourished in a separate but parallel life. This other-William could survive the impatience of waiting to be old enough and the certain return of young Mr Smith, who was, even now, safe from trench war, recuperating from nervous prostration in a French hospital, and from his letters, homesick for the anonymity of his back room.

The impetus to throw it all in and go on the road, go anywhere, quickened under the daily pressure of young Mr Smith's bitter resumption of dispensary duties. Smith, having seen another world and pined for the old one, was disappointed by his countrymen's impatience to get over the war and get on with the new, whatever that may be, and Smith's growing dependency on a self-prescribed, opiate-based pick-me-up mixture created impossible situations in the shop. William was left to take on more responsibility, to cover up, to correct, to anticipate his master's moods, as well as adapt to Dr

Graham's declining faculties and their common patients' loss of faith in medical practice. At home, he faced a parallel decline. His father, formerly a busy and distant real estate speculator, developed a lingering bronchial irritation after the influenza epidemic and was now an invalid who drank in the mornings. The old seemed not to be dying off but rotting away.

William's mother urged him to stick at Smiths'; his sister Jane was of a different mind. Her head full of new ideas, she gave him her Swedenborg to read.

There are some suggestive images in this picture of William's apprenticeship. Forced to begin in a claustrophobic working space under the tutelage of young Mr Smith, instructed to subjugate his personality and talents to the ethics of the perfect product, he was finally 'released' (William's own words, apparently) from the dark cave by a magical old man. Put this way, the old man's initiation becomes allegorical in its impact. A fault line cracked through the middle of what should have been a classical training in pharmacy.

What had begun as the pursuit of a clear-cut unambiguous discipline now unravelled into multiple strands. Jane's contribution, a sort of official sanctioning of the power of individual spirit via Swedenborg, though not transformational (she said he never referred to Swedenborg in later life), injected energy into his plans to create a practice where the man and the medicine could deal directly with the patient. In other words, the domain of the faith healer — a grandiose, self-promoting and occasionally dangerous world, with no code of conduct, no board, no inspectors.

Instead of taking his final examinations, as his family and masters expected him to, William cut himself loose from respectability to freewheel towards another destiny. In the early months of 1920 he turned himself into Dr William Macbeth, late of Pennsylvania, USA. He'd seen *Macbeth* in a Sydney playhouse with Jane and knew the authority of a strong name; *Pennsylvania* put his credentials beyond reach and checking. Dressed the part, with calling cards, Dr Graham's medical bag and a few basic pieces of medical equipment he cruised into the gaps left by the war.

Not a lot is known of those early days except that he practised hands-on healing and sold his own elixirs, and that he moved around country New South Wales with surprising ease, smart enough not to compete with any real doctor and limiting his patient contacts to small acts of faith. The boy who'd pressed his face against the window of Smiths' emerged from his formative years as Dr William Macbeth, well-dressed poisoner.

⊳─┼─◀▸─○─◂▸─┼─◁

The procession of images — William as boy, apprentice, faith healer, husband, father, patent medicine pedlar — follows a certain sort of logic. We all start out on one or another path, wander away and come back. Try on identities.

Seventy years after 1927, I met William in the pages of his sister's diary, the shock of his crime written up as common tragedy: 'Paddy dead. Will leaving Katoomba.' Then a notation of his divorce. Then longer and longer silences, until a reconciliation days before he died from tuberculosis, in his forty-ninth year. In death as in life women fought over him, the winners in this contest being Jane and their aged mother, who buried him according to their wishes, at Botany. The other claimant, the so-called 'nurse for his needs' who shared the last year of his life, was firmly pushed out of the picture.

The missing image in this procession is the one that can never be produced — William mixing strychnine to dose his sons. Like Crippen dosing Cora, it's a concocted scene; it belongs to the world of the culinary master who, presented with sugar and almonds, makes a sublime praline. The concocted image features strongly in poison stories. The space where the picture ought to be, like the abhorred vacuum of nature, gets a hasty treatment from the decorators. However much it disobeys the order to fit in with the rest, the imagined picture has its own integrity and unique landscape and, by virtue of being *non sequitur,* the new construction supplies the some-thing-to-work-on missing bit, on which lawyers and scientists fasten their expertise.

Secrets undermine us. Into the silence around that spare entry, 'Paddy dead', the series of shocks, death and divorce, I fit my own case. Suppose, in the hermetically sealed dispensary William made in Katoomba, forced to labour over his tonics, hating the product yet compelled by training to perfect and polish, suppose the old hate of slaving in young Mr Smith's back room came snarling out of the past to bite him? And surrounding the hate, the fresh pain of new calamity?

Chapter Twelve

One man's meat

THE MEAT AND POISON of William's story is strychnine, the poison nut with bitter seeds. In its benign form strychnine is a useful tonic and stimulant. When it slides out of this safe category through abuse or accidental misuse its full expression is lethal. Once in a long career in pharmacy I saw a strychnine near-miss. There used to be a laxative made by Parke-Davis called Alophen that contained a small quantity of strychnine, barely enough in each tablet to register on a weighing scale. I was in my internship year, gaining experience in community pharmacy, when a three-year-old boy swallowed a whole bottle of Alophens. He quickly became pale then threw up. While we waited for the ambulance I knelt on hands and knees scrutinising the rejected stomach contents, counting the small reddish tablets with their haloes of partially melted chocolate dotted over the mucousy mass. We could account for all the tablets and the boy suffered no more than a ride to hospital, yet the implications of what might have been have always resonated. On calculation, thirty tablets equalled a lethal dose yet the boy did not succumb. Just prior to coming into the pharmacy he'd finished an ice-cream. The amount of time between the ice-cream and the overdose was five minutes. By such small absorption impediments is a life saved. Today, looking back, it makes me think of three-year-old Patrick Macbeth, whose dose of strychnine went straight to the spot. I see in the modern vignette a replay of an event that could have ended happily, a sort of 'sliding doors' look at what might have been if the right action had been taken.

Toxicity is selective. One boy dies, another lives.

———————

At the level of the organism we know the numbers that relate to how toxic a substance is through a measurement called the LD_{50}, or the dose killing fifty per cent of test animals. At a cellular level, a toxic agent may injure certain cells without harming others, sometimes cells which are close neighbours. Across species we can say things like: man is fifteen times more sensitive to atropine than the rabbit, cows feed on the deadly nightshade with impunity and pigeons are almost entirely unaffected by opium. How such comparisons come to be known owes something to folk wisdom and a lot to the labours of toxicologists. (How else would we know *not only* that the rabbit is immune to the Death Cap mushroom *but also* that the reason lies in muscarinic activity.)

The human capacity to flout measurements like the LD_{50} is apparent right through the literature of toxicity.

In the nineteenth century when *arsenophagists*, or people with an arsenic habit, were most prevalent, James Maybrick used to stop at a Liverpool chemist shop up to five times a day for a dose of arsenic solution. When he went abroad, which was often, he took a supply of doses with him. Maybrick famously died from an overdose of arsenic in 1889 and his wife Florence, always protesting her innocence, was convicted of murdering him. In the trial reports we read of him boasting to friends who watched him stir poison into his food, 'I take this arsenic once in a while because I find it strengthens me.'

Opium-eaters were another anomalous group. 'Almost incredible quantities have been consumed by such persons,' wrote Dr Blyth in 1884, 'and yet a high degree of health, both physical and mental, is enjoyed.' Thomas de Quincey's *Confessions of an English Opium Eater* (1822), gave the world its first literary exposure to the interior life of an addict whose own peculiar dreaming faculty, raised to a higher order through the mediation of opium, allowed him to make (in a pre-Freudian age) connections between lived experience and the symbolic language of dreams. De Quincey took his opium as laudanum, an alcohol-based drink which was freely available, cheap to buy, and socially unremarkable in normal levels of usage. Where he deviated from the commonplace practice of a few drops for headache or insomnia, or even the once-a-week binge-drinking of workers who couldn't afford gin, was the rapid escalation of his needs: from monthly use, to weekly, to daily, to a staggering 8000 drops a day, which he claims to have exceeded at the zenith of his engagement with 'the dread swell and agitation of the storm'. De Quincey used opium for more than fifty years, the latter part of his life on reduced doses, with at least one relapse into uncontrolled addiction.

In 1875 at an inquest held in Chelsea on the body of an infant boy, it was ascertained that the mother had given the child 'poppy tea', an infusion of two poppy heads in hot water, and that the boy, after drinking several bottles, had fallen into a deep sleep and died 'with all the symptoms of narcotic poisoning'. The 'pernicious nineteenth century practice' (Blyth's words again) of giving infants various forms of home-brew opium, or proprietary brands sold as *Soothing syrup, Nurse's drops,* or *Infant's friends,* to settle restlessness, and to keep them — for what seemed to be the greater part of their existence — *asleep,* has been legislated out of existence, only to resurface in our times in disguised forms.[33]

><+>+0+<+>+<

The month the name puzzle was solved, I ordered Patrick's death certificate. Easy once you have the right surname. For the first time since 1980, I had document corroboration. 'Death by strychnine, innocently self-administered. Signed. *A. Judges, Coroner.*' But why deny a coroner's report? Because of that phrase 'innocently self-administered'? Isn't the verdict a triumph, doesn't it imbue the story of the deaths with aesthetic sense? In a short space of time, three new voices had been added to Rose's: Jane, the indulgent sister; the coroner; and William's own words lifted from Jane's diaries. A chorus instead of a solo.

Rose's testimony has its own, defensive context. To a farm girl who'd led a trimmed down, hard-edged life, in and out of various forms of deprivation, where doctoring fell within the domain of the average sensible woman, and most city people were automatically suspect, a man with pretensions to a style and set of standards irritated her sore spots.

⊷⊶⊷○⊶⊷

William Macbeth had played at being a doctor in childhood, as many of us do, but unlike the majority of young dreamers who are snapped out of it by quotidian realities, he lived out his relatively short adult life in a sustained yet borrowed reality. What began as a trick of simultaneity became, through practice, a habit and then a compulsion. Having taken the surname Macbeth in 1921 to lend credence to *Macbeth's Tonic*, an interim identity change for the duration of his faith healing tour of country New South Wales, it stuck and couldn't be thrown off without embarrassing disclosures. Inevitably some of what had happened during William's apprenticeship affected him more lastingly than other experiences. The contribution of old Mr Smith is significant. Just as transformational was the *de facto* apprenticeship he served with Dr Graham, who through failing memory and faculties, brought the young man ever closer into the privacy of the doctor–patient interaction. This invitation to the inner sanctum is seductive to a receptive personality, a bit like listening at the door of the confessional, and being told: It's all right, I want you to hear. (I blush to think of my own experience at deliberate eavesdropping. Long ago when I was stuck in the back room of an old pharmacy compounding medicines in a trance of boredom, I put a glass to the common wall between the dispensary and the surgery next door to better hear the muffled voices that hummed off and on all day. Most of what I heard was unremarkable, but anything overheard has the potential to violate some taboo, and is therefore magnetising.)

William absorbed the influences of his two mentors, old Mr Smith and Dr Graham, so fully into his interior landscape that he became at first one then the other, then finally a blend of the two where all traces of the originals had melted away.

After his faith healing tour of 1922 he successfully reinvented himself at least three times. Once as a dentist in a country town, once as an evangelist minister, and, most successfully of all in a financial sense, as an attendant psychiatrist in a mental institution. William's impersonating self seems to have lain dormant like an opportunistic virus waiting for the right conditions to nourish it into life again. In the 1930s, drinking heavily, separated from his wife and surviving sons, temporarily directionless, he signed himself into a sanatorium for a rest cure. In the quiet gardens of the hospital he sat down beside Gordon Macauley, a true Scotsman of the same age as himself, and struck up a conversation. In an accident almost designed to activate William's *alter ego*, Macauley mistook William for the attending physician. Macauley had an inheritance and a full-blown delusional psychosis. When Macbeth and Macauley left the sanatorium together, Macauley believed that William had signed his release papers and was suspending his wider practice to devote exclusive attention to Macauley's recovery. For his troubles, William received two cheques, one for £3000, another for £5000 a month later. Macauley inscribed a Bible to 'Dr Macbeth, fellow countryman, brother in Christ, one who has proved himself worthy in the sight of God.'

William bought a car, employed a chauffeur and took up clay pigeon shooting.

Chapter Thirteen

Secundum artem

ABBREVIATIONS USED IN PRESCRIPTIONS
S.A. SECUNDUM ARTEM, ACCORDING TO ART,
I.E. WITH PHARMACEUTICAL SKILL

Pharmaceutical Handbook, 1970

WHEN GUSTAVE FLAUBERT published *Madame Bovary*, the pharmacists of his region met to discuss calling on him at home and slapping his face.

With good reason. The fictional chemist in the book, Monsieur Homais, is a pompous man who might have been a minor and easily dismissed character in a novel full of stronger characters, but is given a major role in the unfolding drama of Emma Bovary's death, and also the final triumphant sentence of the final chapter. Flaubert spent twenty years inventing Homais, the embodiment of everything he found bourgeois and ludicrous in French society. Homais spoke in clichés, had a brash laugh, and habitually inflated his own importance. The chemist was so alive to the author that he quoted his sayings in letters to his friends, writing, 'As my pharmacist would say . . .'

To be fair, chemists were not the only victims of the author's exquisite dissection of French bourgeois society; medical officers with a rank lower than surgeon (like Emma's husband) were shown to be little more than educated fools. The fact that Flaubert's father had been a surgeon was not lost on the outraged pharmacists, or the local medical fraternity.

><+<>+<>+O+<>+<+><

Squabbling between chemists and doctors has its roots in medieval times and its flowering head in the nineteenth century. Some time during the Middle Ages a medical hierarchy worked itself into being. At first physicians sat squarely at the top. Surgeons climbed onto a lower rung of the ladder. Apothecaries had one foot on the ground and one on the bottom rung, and all the rest, the midwives, healers, nostrum sellers, bone setters, tooth pullers, herbalists and other ratbags were vilified into obscurity. English chemists, who were once members of the Company of Grocers, united under the 'Masters, Wardens, and Society of the Art and Mystery of the Apothecaries of the City of London' more than a century before English surgeons shook themselves free from the company of barbers. Once both groups had a voice and a membership, they set about obtaining parliamentary approval to protect their territories. It was only in 1858, just after the publication of Madame Bovary, that England passed an Act dissociating medical doctors from 'pill compounders'.

One of Monsieur Homais' sins was to practise 'innocuous consultations' in his back room, a habit he should have given up when a real doctor came to town. This habit sits like a thorn in the foot of medicine's quarrel with pharmacy; it throbs at the centre of the power dynamic between those who prescribe and those who dispense. A third, often unacknowledged, force in this relationship (a force that turns it from a tug of war into a triangle) is the patient who may exercise choice no matter where parliamentary acts draw their lines. The patient develops one relationship with the doctor and another with the chemist, and like a woman with two lovers, pledges loyalty to both.

The demarcation lines blur when there are more sick people than practitioners to serve them, and where topography and distance are hostile. In some of the most surprising First World locations, the chemist is still the poor man's doctor; and isolated doctors still have their own dispensaries. The line blurring, to my mind, comes from the scuffling of patients' feet as they tramp from one group to another on health pilgrimages. Anyone who has been ill, and has faithfully taken his prescription from doctor to chemist, swallowed the remedy yet remained ill knows the futility and sense of abandonment that can come with the exercise.

To some extent advertisers have stepped into this breach. We browse the shelves reading labels that promise cures, investing the anonymous words of a spin-doctoring marketing team with the trust we used to give medically trained professionals. Not all of us, but enough to worry the establishment into occasionally issuing denunciations of the latest 'cure'.

The ready-made market, now dominated by multinationals, was once the secure niche of the quack who dealt not so much in self-limiting conditions, but in miracles. With a flourish of his silk handkerchief, the quack pulled out a cure for cancer, impotence, venereal disease, and the thing we all lose, youth. 'Stop dragging along. If your household or business cares exact so much energy that sometimes you feel run down, it is wise at once to take this product.' 'A strong Manhood and Womanhood is your birthright. It is Nature's gift, and if it has been impaired or dissipated, now is the proper time to regain it. Take this product, the Life Giver. Quarts: five shillings and six pence.'

My grandfather was a chemist who posed as a doctor in country towns. I met a couple of his patients in 1980. Macbeth was a good bloke, they said, because he listened, and he liked a drink (this told with wry laughs) and would stand a round at the bar. Did he write prescriptions? I asked. Didn't bother with them, came the answer. Made his own stuff. Used his hands a lot. Nice and warm, they were.

Therapeutic touch and patent medicines, the ratbag end of medical practice, still don't amount to serious breaches of professional care if that's as far as the trans-gressions go. But in the last years of his practice William pulled teeth, cut and sutured skin, and set bones. The question I ask is, does a man with no appreciation of boundaries (legal, ethical, or professional) manifest reckless behaviour from a lack of structure in his own interiority? Does he inhabit a world that is nebulous and disorganised, does he drift and dream free from the moorings that hold the rest of us anchored in the real world? If this is a picture of William at work, then I have to consider the negative expression of a lack of bound-aries: that is, an ego that believes itself to be God-like, omniscient, and capable of anything.

Modern chemists are regularly consulted for advice. The dynamics of these free, unscheduled exchanges are well understood by chemist and patient. In busy city pharmacies, both parties begin and end the transaction as strangers; something is asked, a response is offered, and both parties often return to their pre-consultation position with no intention of carrying it any further. Busy medical centres may also, I'm told, induce this utilitarian dynamic in the doctor–patient relationship. What happens to this managed, almost democratic, arrangement when one of the parties crosses the line? When, for instance, a chemist prescribes, a doctor misdiagnoses, or a patient misrepresents herself for the purpose of obtaining drugs? The response, unlike the days of Homais and William Macbeth, is swift and punitive. As knowledge becomes accessible to all, as power balances shift, as regulatory bodies and legal statutes tighten up any loose connections, we are, none of us, in any doubt as to who may do what to whom.

><+>+-0-+<+-+<

Nearly every day of my working life in pharmacies I've practised my own brand of back room consultation. I can't judge if the encounters have always been innocuous. As layers of secrecy are peeled off the intention and nature of medicines so my job has changed from typing 'The Tablets: Take one three times a day' (which carries the sense: *and do what you're told*) to printing off screeds of computerised and very detailed pharmacological information for people to take away and worry about at home.

Once when I worked in a country town I threw aside

the computer notes and listened to a woman's story of widowhood, loneliness and psoriasis. In her purse she carried a photograph of her husband in army uniform and we both agreed he'd been a handsome man. Soon she was calling in every day and I began to feel the weight of her neediness and to detect a sort of double-dealing in myself because my stay was coming to a close and by encouraging her, I was setting myself up to be just another person who befriended then deserted her. On the final day of my locum period she came to say goodbye and to show me her legs. She lifted her skirt so I could see where the skin was improving. I ran my hand over the inflamed scaly patch we'd been watching together for weeks and we talked about taking care of ourselves.

Suddenly she said, 'You're the only person who's touched me since Ed died, I mean actually put a hand on me.' Like the lamenting party left behind in a love affair, she wished me well in the future.

The chemist who supervised part of my early training was a failed doctor. He'd switched to pharmacy as a poor second (after failing his exams), bringing all the contempt of someone with a higher calling to the poison-vendor's profession. I met him again in Romeo's lines, 'I do remember an apothecary/ sharp misery had worn him to the bones', and in the description of William's master chemist, young Mr Smith, that bloodless, disappointed man.

Like Homais, the chemist who trained me had pretensions. One of them was to give me instructions in Latin, the sharpest weapon he could find to cut me down to size and debase our pupil–master relationship into a power struggle. My only outlet, being too cowed to speak up, was to indulge my mind with revenge fantasies, nearly all involving poison, so it was with piqued interest that I read the section of Agatha Christie's biography dealing with her early instruction as a trained assistant in my profession.[34]

During World War I, young Agatha began work in the newly created dispensary of a Torquay hospital. She was to assist the incumbents and study for her Apothecaries Hall examination. In a short time she was mastering the Periodic Table, learning atomic weights and churning out jars of ointment. She also practised the Marsh test for arsenic in the Cona coffee machine (and blew it up in the process). She spent Sundays in a community pharmacy for further training under Mr P, an excellent, if too familiar chemist, who showed Agatha a 'dark-coloured lump' he carried in his pocket. He asked her to guess what it was then surprised her with the answer 'curare', harmless by mouth, deadly on the tip of an arrow. The formerly trusting young Agatha viewed her employer differently when he explained that he kept curare in his pocket to make himself feel powerful.

During lulls in her dispensing work Christie plotted detective stories. Poison was perfect for what Christie called 'the *intime* murder'. And she never forgot Mr P and his lump of curare. Nearly fifty years later she resurrected him as Zachariah Osborne, the thallium poisoning pharmacist of *The Pale Horse*.

Primo Levi, another kind of chemist, studied his subject for the laws he believed would be revealed by the unravelling of its mysteries — principles of order in himself and the world around him.[35] On any line of a chemist's path through a professional life, he says, he should be aware that 'on one of those pages, perhaps in a single line, formula, or word, his future is written in indecipherable characters, which, however, will become clear "afterward".'

While still young (eleven, in fact) he had tangled with hydrogen gas. He put a match to an outlet and was shaken by a small, angry explosion. It left him with trembling legs and a confirmation that he had tapped into a force of nature. Hydrogen, the element at the angry heart of the sun and stars, 'from whose condensation the universes are formed in eternal silence', taught the young chemist who would almost end his days in Auschwitz valuable respect for the invisible world.

Christie and Levi, though galaxies apart in type and artistic expression, were both impelled by some inner force to write about the mystifications of chemistry. Levi's story of the bootmaker who suspected his sugar had been poisoned with arsenic is a small masterpiece of eloquent narrative.

These and other writers with special knowledge worked at the popular, audience-oriented end of a continuum; at the other end were pure researchers also seeking to make sense of the hidden power of chemical bonding. The writing chemist, and doctor, contributed invaluably to the sum total of our knowledge of chemicals. Processing, examining, reflecting through the printed word, all laid down trackable spoor on the path to understanding. Modern forensic toxicology grew into the elaborate and exact science it is today on the foundations of a few writing chemists who plunged their hands into the bellies of poison corpses and extracted the contents in the hope of finding some sign. They used crude tests. If you threw powder residue onto a red hot iron and it smelt like garlic, arsenic was your villain. As late as the 1880s it was common practice when faced with a poison, to taste it and compare it to known samples from a collection. It wasn't long before *post mortem* stomach contents (and for that matter, any matter ejected before death) became the star witness at an investigation, and why a serial poisoner like the notorious Dr Palmer made such a serious effort to sabotage his last victim's autopsy, jostling the stomach so the contents spilled, then bribing the courier to shake the sample thoroughly on its journey to London.

Forensic medicine as a discipline was unheard of before 1918. In some countries dissection of bodies was considered outrageous; in America for instance it was only legislated for after the Civil War. There was no co-ordination between the pathologist, the chemist, the serologist; no template for processing body fluids, and no coroner to report to.

The problem with stomach contents was finding the poison in the partially digested mass of solids and liquids. In 1836 James Marsh sorted it out, developing his test (the same one young Agatha attempted) until finally it was possible to separate arsenic from soups and ales and the mish-mash of stomach contents, and forensic certainties could stand against personal suspicion and superstitions.

It may be that the eventual synthesis of a working system for tracing poisons all the way from the hand of the poisoner to the belly of the victim grew out of the power struggles of disparate groups clamouring to have their voices heard.

Chapter Fourteen

Dry white arsenic

THE QUEEN OF POISONS, arsenic, is a defeated monarch at the end of the twentieth century. All her splendours, sorrows and tragedies are far behind her. The old lady has been banished to a grace-and-favour apartment in the grounds of the palace; remembered, even revered, but rarely seen abroad any more.

In another era, arsenic had a glittering reputation as a deadly, but slow, poison. Used carefully, arsenic the death-bringer could also be of practical use. The Romans cleaned festering wounds with it, mission doctors slapped it on leprosy and syphilitic sores, nineteenth-century ladies whitened their complexions by bathing in it, and the horse tamers of the Tyrol crunched great hunks of it between their teeth 'for virility'. Both praise and blame attached to its reputation; whatever the outcome arsenic could never be ignored. Men of science wrote like courtiers with cases to plead: 'Its poisonous properties are known to everyone and its toxicology overshadows that of all other poisons in importance.'

Only a minority of people know what arsenic looks like any more. The arsenic catalogue, once a poisoner's delight, is largely redundant. Metallic arsenic is a subterranean dweller of largely uninspiring appearance: greyish, dull and reclusive. But through marriages, interbreeding and chance encounters with other elements, it forms compounds as colourful as a new spring collection. The ruby red crystals of realgar, found in nature or made commercially (by heating arsenic with sulphur) were once prized by pyrotechnic artists for firework spectaculars. An old recipe for producing a blue fire is sulphur, nitre, antimony, realgar and charcoal. A brick-red arsenic results from mixing arsenic with iodine.

When arsenic is combined with slightly less sulphur (60:40) a brilliant yellow powder is produced, the King's Yellow or orpiment, used by illustrators and decorators on projects like the Book of Kells and the Taj Mahal.

The blue form of arsenic is a mixture of metal with copper and potash, while the greens (Schweinfurt and Scheele's) are variable combinations with copper alone.

The pure element, number 33 in the Periodic Table, chemical symbol *As*, is almost never seen, even in the earth's crust where it originates, because arsenic has such a tendency to make friends. When ores are mined, the arsenic comes to the surface and is driven off in the smelting process; sometimes a mineral spring or geothermal power plant floats it into the upper world. Common white arsenic is a crystalline white cake.

The rarer gas form, arsine (colourless and non-irritating though it has a distinctive and foetid garlicky odour), represents arsenic at its most lethal. The gaseous form of arsenic is a chemical accident. It has no industrial or commercial use, it comes into being when metals like zinc are treated with hydrochloric acid, or when arsenic wallpapers and decorative mouldings are attacked by bacteria, as in the famous case of the American ambassador to Rome who slept under a ceiling of peeling arsenic roses. So toxic is acute arsine poisoning that it can only be treated with blood transfusions, oxygen inhalation, or agents to halt the destruction of red blood cells.

The lethal quantity of white arsenic varies from individual to individual but is generally accepted to be in the order of 100 to 300 milligrams, slightly more than what would fit on the end of a dry, pointed knife dipped into the powder. You can take your arsenic in almost anything: bread and butter, porridge, cheese, blancmange, meat juice, cocoa, coffee, soup, dumplings, elderberry wine, light beer. No one has yet produced the definitive list.

Arsenic in the mouth tastes sharp, like a burning sting. Mixed with saliva and swallowed it descends to the stomach, that chamber of the body programmed to deal with strange and texturally diverse brews. A healthy stomach can cope with bacon and eggs, kippers and tomato sauce, toast and marmalade and three cups of strong tea but it baulks at a shipment of poison dumped on its wharves. If we're lucky we vomit courtesy of the gag reflex and reverse peristalsis. What went down comes back up, with force. In very lucky cases arsenic gets no further than the mouth. The salivary glands respond by rapidly diluting the dose, and fear makes us spit.

Industry uses vast quantities of arsenic. The white oxide goes into smelting and refining of ores, as well as taxidermy, and the tanning of skins, furs and hides, while the coloured varieties, green, red and yellow, are ingredients of pigments used for wallpapers, paints, prints and aniline dyes. In 1958 the world production of arsenic was

40,000 tons and this increased to 55,000 in 1962. However, by the 1970s, after the organic pesticides became freely available, the demand for arsenates fell by more than fifty per cent.

A healthy (read 'non-poisoned') person of average height and weight holds about 18 milligrams of arsenic in their cells, muscle, sinew, bone and nervous tissue *as a given*. It gets there through our lifelong interaction with the environment — through the soil, the sea, the air. Five parts of every million parts of the earth's crust are arsenical. These dangerous molecules go round and round in a cycle, so it's not surprising to learn that each body has a certain tolerance level to arsenic. Forensic pathologists have to take this level into account when they examine *post mortem* organs for evidence of contamination. We eat poison voluntarily, involuntarily, and sometimes we know about it but we don't care. Arsenic is, whether we like it or not, a resident alien in the body's citadel.

>‑<>‑O‑<>‑<

Arsenic-eating was once a European family pastime. The *British Medical Journal* first reported the practice to a fascinated English audience in the 1860s although the practice was much older. On the hillsides of the Tyrol, in Hartburg and other districts, the local stablemen perfected a technique that was passed on from father to son. The initiate began with a millet-sized piece of white arsenic — readily available on the slag heaps of ore smelters — and progressed to a piece the size of a pea. In Hartburg, the phases of the moon played a role in eating rituals. At the new moon, the dose was small; at full moon the dose might reach as much 4 or 5 grains.

It began, apparently, with horses. 'In all European countries grooms and horse-dealers are acquainted with the fact that a little arsenic given daily in the corn improves the coat. When a horse has been dosed for a long time, it seems necessary to continue the practice; if this is not done the horse rapidly loses condition.' In Hartburg it was given in large quantities, as much as 50 to 100 grains per day in incremental doses. An arsenic-fed horse became vivacious and excited. Unscrupulous horse-handlers the world over have used arsenic as a 'coat-shiner' in lieu of elbow grease, despite the threat of prosecution under the Poisons Act.

A little bit of what turned the horses on was appropriated for human use. These mountain men were observed, by the medical professors who examined them, to be robust, vigorous and long-lived, if a little puffy around the face. Arsenophagists, as they were called (there were still a few left in 1978), became in time embarrassingly fat. 'Not a rounded over-fed fatness, but blown-up, unnaturally bloated creatures whose long acquaintance with poison had produced a dependence identical to any better-known chemical dependence.' The arsenic-eaters scoffed at doctors who warned against the practice. One man crunched a large piece between his teeth in front of a visiting English doctor, saying he ate that amount three to four times a week. Many homes in the area were found to contain supplies of white arsenic which were freely available to wives.

As a group, the arsenic-eaters enjoyed their status. They particularly enjoyed being the subject of scientific and public scrutiny. Without an audience of cowards and sceptics, their perilous behaviour might have lost its shock value and subsided into boring eccentricity.

Inevitably, when word gets around, a queue of thrillseekers will line up behind a new idea. By the 1870s, Englishmen and Americans had acquired the arsenic habit. It was relatively simple to obtain an arsenic drink, or pill, from a chemist. Ostensibly, arsenic prescriptions were issued for symptoms of malaria and unspecified conditions like debilitation, in reality the target (for some) was sexual potency. '*Arsenikon*', from the Greek, means *potent*. By 1882 the habit was being reported in provincial newspapers: 'We are all perfectly aware that men-about-town are as much in the habit of taking these dangerous drugs, strychnine and arsenic and what-not, as they are of drinking champagne and smoking tobacco.' Elsewhere, the *British Pharmacopoeia* stated its laconic position: 'There is no evidence that arsenic directly causes an increased formation of red blood cells in normal individuals.'

Dry white arsenic sold over the counter was routinely coloured with soot or indigo in the days of lady poisoners and gaslight thrillers. Even though forensic science was in its infancy, the methods were critical enough to discern dye-stuff in stomach contents. Madeleine Smith, the well-known nineteen-year-old Glasgow girl who allegedly poisoned her French lover in 1857, purchased arsenic several times from a local pharmacy, giving 'rat extermination' as her reason in the dispensary register. Madeleine's arsenic did not, as far as anyone could tell, end the life of any rodents. She used it cosmetically, to whiten her skin, a practice said to be one of the vanities of the nineteenth and early twentieth centuries but which almost certainly dated back to Elizabeth I's reign and beyond. At her trial, the apparently obvious source of the arsenic was brought into doubt when the dead lover's habit of taking arsenic drinks himself (also for his complexion, but more probably for the same reason as his Tyrolean neighbours, that is, as an aphrodisiac) was brought out. The deceased's organs contained no dyed arsenic, but they did contain a vast quantity of poison, nearly thirty times the lethal dose, which, it was argued, would be inconsistent with intentional poisoning. These two clinchers got Madeleine off with a 'not proven' verdict. She died in America aged ninety-one.

Arsenic has often kept bad company. Royalty and commoners have sought it out with criminal intention; individuals and vendors have promoted it as a sex aid; and quacks have blended it into fanciful nostrums for miracle cures. Pills made from arsenic and ground pepper once circulated as pick-me-ups. Powders of arsenic, hemlock, cinnabar and 'dragon's blood' cured everything from cancer to rickets.

The context for deliberate arsenic poisoning is domestic. The motive may be political, religious, personal or even philosophical but the action usually begins in a kitchen. Lethally precise doses can be easily mixed at the sink and carried to the sickroom without arousing suspicion. Soups and stews have served arsenic poisoners well. One serial killer from the 1920s served her sister-in-law arsenic-laced soup (which the cook and the cat had tasted and *both* been violently sick from) without implicating herself in front of two attending physicians.

Chapter Fifteen

Napoleon's hair

A CONTROVERSIAL CASE for murder by arsenic was made jointly by three doctors in 1961. Their subject was Napoleon I, said to have died of 'stomach cancer' on 6 May 1821, while exiled on St Helena.

The cancer story had been disputed from the start, principally by Napoleon's own physician, who diagnosed longstanding severe hepatitis.

What alerted the modern doctors to arsenic was the non-linear progression of Napoleon's symptoms. The horrors came and went. He was worse when being cared for by English doctors, then improved when he was moved on board a ship, or into a simple cottage with a friend and his son.

When the attacks came back (coinciding with the reappearance of an English doctor, so it was said) they lasted a week, and his legs collapsed under him. His swollen feet were permanently cold, his teeth ached, he coughed, had liver pains, developed pimples around his mouth and generally looked as if he were about to die.

From March 1821 his gums bled, his teeth loosened, his sight faded, he had a burning thirst and a coated tongue. The three doctors who looked at the case studied the ups and downs as an entity rather than as a sequence of unrelated and overlapping illnesses. They deduced that Napoleon had been suffering from longstanding slow arsenic poisoning, interrupted by bouts of short sharp doses of the poison.

Arsenic leaves its calling card in hair shafts. It can be measured in the hair roots within half an hour of ingestion. The technology to read the deposition of poison in hair, like layers in sedimentary rock, was perfected in the 1960s just as arsenic was beginning to disappear off the scene. By an extraordinary bit of luck, the scientists had access to a family heirloom — a bundle of genuine, authenticated hairs taken from Napoleon's head, knotted and attached to a piece of paper. They set to work with a nuclear reactor and Geiger counter to track and graph the arsenic-16 isotope, and yes, the tests showed intermittent exposure to arsenic, with abnormally high exposure over a four-month period.

The idea was intriguing, but difficult to prove without exhumation and a second autopsy. The hypothesis lies around in the literature like Cleopatra's hair combs, highly suspicious but impossible to verify.[36]

Chapter Sixteen

Death by misadventure

ARSENIC COMES HONESTLY by its bad reputation. Statistics from the *British Medical Journal* were collated out of 1000 recorded cases of poisoning between 1752 and 1911: 43 per cent were homicides, 23 per cent suicides, 20 per cent accidents, 3 per cent deaths from abortions, less than 1 per cent deaths from quack medicines, and 10 per cent unknown. However, reporting in each category is notoriously unreliable and what may look like an accident may be a premeditated murder.

To arsenic's reputation as a sophisticated eliminator of princes and emperors must be added the clumsy handiwork of poor sinners from the lower ranks.

>-+◆>-O-◆+-<

In the literature I found John Erpenstein, a Prussian immigrant living in America in the nineteenth century.[37] For a few days in 1852 he achieved a salacious sort of notoriety as the man who poisoned his wife with an arsenic sandwich so he could marry his mistress.

Nothing in Erpenstein's former life appears to have primed him for fame, and we know a lot about his background because he repented in a lengthy document published after he was hanged. He learnt the tailor's trade in Germany and at twenty-four married Fritzigen, a poor Protestant girl. Years passed, a war intervened, then in middle life he heard the call of the New World and sailed for New York to make the family's fortune. Fritzigen and the children stayed behind. At first she encouraged him to be the family pioneer, then grew doubtful about moving the family so far from everything they knew. After some time she refused to budge.

Erpenstein wrote, pleaded, sent money and asked others to intercede. He'd set himself up as a tailor in New York and could see the possibilities to get ahead.

Around this time he employed Dora the seamstress. Dora was young, Erpenstein was lonely. He began buying her presents, taking her out, and eventually sleeping with her. Her mother and sister approved, believing I suppose that he was single or widowed, or somehow parted from his wife. Erpenstein promised the girl marriage.

Their affair was just rising to the boil when Fritzigen changed her mind, booked a passage on a ship and set sail for New York with the children. Simultaneously, Dora announced she was pregnant.

'I felt as if between two fires,' he wrote in his confession, 'and matters grew steadily worse.'

The two women, understandably, hated each other at first sight.

Erpenstein had his razor sharpened so he could cut his throat but couldn't go through with it. 'I reflected on what course I should pursue. Something suggested *get arsenic*.'

This ordinary, rather foolish man confronted the reservoir of dark thoughts and deeds we all carry, saw a solution that seemed final and achievable, and embraced it. At the German chemist shop he bought a quantity of powder.

Unfamiliar with poison, Erpenstein tried a portion on himself. No sooner had he swallowed a small amount than he vomited with no consequent ill effects. Troubled, he reconsidered his method. Perhaps he'd been tricked by the chemist into buying a harmless white powder. He tried again, increasing the amount of poison, and again he vomited. For some reason, the arsenic, if it was arsenic, would not take on him.

There was no stand-by plan. He was committed (by his inability to think of something better) to the poison, so one day he made a sandwich for them to eat while walking to the city. He sprinkled arsenic onto a slice of bread, spread butter over the top, then doubled the slice over. Halfway to their destination his wife wanted to eat. She put her hand into his pocket and took out the bread. In a state of rising panic, Erpenstein suggested they halve the slice, with the crust to go to him. He stated in his confession that at this point his half 'fell into the water' while his wife ate hers.

Fritzigen did not throw up. She swallowed her bread and struck out on the final mile of their journey. Her husband's thoughts raced from possibility to possibility — was she poisoned, was she immune? How long before he saw a result? She had taken so little of the arsenic, perhaps nothing would happen at all, and then what? At their destination he bought his wife two cakes and two apples, which she ate and then became violently sick. 'I supported her in my hands. My heart warmed towards her —' Eventually they returned home to endure her last miserable hours together, Erpenstein weeping loudly as his wife groaned.

How could Frau Erpenstein walk many miles then eat cakes and apples after her poisoned sandwich? Fritzigen's strong constitution might supply part of the answer; the rest is probably the work of the bread-and-butter coating. Presuming that in her hunger she swallowed without too much chewing, her part of the sandwich might have descended to the stomach in a securely wrapped parcel which had to be dissolved before its awful contents were let loose.

Erpenstein himself, like Emma Bovary, put raw arsenic in his mouth. Emma 'crammed' in a handful and it 'took'; Erpenstein coaxed his dose off the end of a knife and it came straight back up again.

There are hundreds of stories to contradict the smooth chain of events laid out by textbooks (that is, that following a lethal dose symptoms appear in an orderly sequence). One, recorded by Alfred Swayne Taylor, a pioneer of toxicology, tells of a victim who remained healthy for eight hours after a toxic dose then suddenly died. The post-mortem showed a cyst on the stomach wall covering the arsenic, a freakish protective growth that captured and held back the poison, before bursting under biodynamic pressure. One man died after eleven days of gradually developing symptoms, the most prominent of which was acute sensitivity to touch.

❦

Erpenstein's confession, published as 'A Poor Sinner's Life, and His Guidance, under God, to the Lord Jesus

Christ', is a wearying ramble through his early years, with the occasional surprise of a perfect sentence laid out on the page. ('My mother was a toll-gatherer's daughter' convincingly endears the man to me if not to history.)

Just days before his execution, Erpenstein was visited by 'a good Christian woman' in company with a priest.

> To my great astonishment she brought out a Bible, entirely new and furnished with a silk mark, and handing it to me, requested that I should make therein some family records for my children. I resolved never to cease praising God for the kindness he had shown me as a Father in Heaven, in sending so rich a lady to visit me, and kindly take my hand and minister consolation to me.

The murder of his wife is just one more chapter in a life of events — usually involving women and usually beyond his control; he gives equal weight to the seduction of Dora, the murder of Fritzigen, the nice Christian lady with her Bible, and his final sweaty reconciliation ('the perspiration rolled from me in my agony') with the Catholic faith, for which he is posthumously applauded in an afterword by his priest.

William Macbeth's life didn't end at the gallows. He lived for twenty years after his sons died, in fluctuating circumstances, and in an almost constant state of re-invention. This outcome might be read as a triumph of audacity, or an accident of statistics. The mis-catalogued deaths of two small boys will never bend the numbers out of shape. Perhaps it lends some weight to the argument that men who successfully poison work inside the 'industry' that supports its existence. Those who stumble in the side door, like Erpenstein, suffer the fate of heavy-footed interlopers.

><->-0-<+-><

True accidental poisoning, the statistics say, happens about one in five times.

In the summer of 1925 a young Australian school-teacher died after an evening of 'cruel pains in her stomach', consistent, ventured the attending doctor, with some kind of colic.[38] She'd been suffering pain for days, initially attributing her troubles to the rich country food provided by her billet family, for which she compensated with exercise and increased fluids.

Coincidentally, or otherwise, the farming family had also been ill, but after a short holiday all regained their former good health and no connection was made between their illness and the teacher's.

Connections of another sort, however, were made when the autopsy revealed death by arsenic poisoning. The police were notified, statements taken, and an investigation begun. In common criminal investigations the most useful thing to have is a suspect who has a motive, and preferably the means and opportunity to kill. What the suspect did and what happened to the victim should fit each other the way a glove covers the shape of a hand.

At the mention of poison, the farmer pointed a finger at his neighbour — another vigneron — a rival who'd stop at nothing to disrupt the competition. Clearly, said the farmer, his neighbour had planted poison in the house, but it had missed its mark and taken the life of the pretty young girl sent to the country for her first teaching assignment. This unsupported assertion was backed up by everyone interviewed, and endorsed by the local constable, who described a history of bad feeling between the two vineyards.

An investigator sent from the nearest capital city was briefed on the background of rivalry and set about collecting evidence to test the assertions. It's at this point in this sort of case history that one catches a glimpse of forensic endeavour serving the strongest interest. The accused neighbour (who at this point didn't know he was the subject of suspicion) becomes a passive object of scrutiny, the X on the map that marks the spot that all tracks lead to. Murdering with poison is, on the whole, a one-man or one-woman exercise, involving step-by-step subterfuge: secret manipulations (mixing the dose out of sight of witnesses), brazen invasion of the victim's privacy (adding the dose to food or medicine) and, as one nineteenth-century writer liked to characterise poisoners, 'the exclusion of any sense of pity'. In the case of the dead schoolteacher, a demon fitting this description lived a mile down the road.

The case takes an interesting turn when the out-of-town investigator turns out be a man who likes to come to his own conclusions. He carefully searched the farmhouse and surrounds for poison and found nothing, not even a bottle of medicine; nor did he find among the family itself or the neighbours (including the suspect) any evidence of psychopathic tendencies, or an unnatural interest in, or knowledge of, chemicals. The dead woman had no disappointed lovers; the school knew her as a conscientious loner finding her feet in a demanding job; the farmers had invited no strangers or drifters inside the house.

The investigator (a pathologist employed by the Crown) moved into the farmhouse to make his job easier. On the surface the case seemed like a clumsy outback murder gone horribly wrong, or, if the girl was the target, exactly to plan.

I should say at this point that I've deliberately chosen the schoolteacher's story (and to tell it in an unfolding, as-it-happened way) to show something of what I've called the 'tidal pull' of description when spectators crowd around the poison vessel. The players in the story — the dead girl, the farmer and his wife, the local constable, the protesting but guilty-looking rival farmer, the headmaster who represents neutral authority, the townspeople — might have come from central casting, and the 'star', the lone gun from outside who rides in to solve the mystery, could be any hard-boiled romantic lead of the day. Melodrama is never very far away when the cry 'poison!' goes up. Screenwriters know what to do with a set-up like this one; each of the players has to reveal a not-immediately-obvious potential for culpability, and the audience has to be taken on a wild and frightening ride to each of the dead ends before, in a surprise climax, the real killer is unmasked.

Screenwriting, thriller-writing, popular journalism all steal their Promethean fire from much older stories, from myths and fables that oppose 'good' and 'bad' characters moving through a story. Anti-climax is not good cinema. Complex unravelling tries the patience of the average audience.

The story of the schoolteacher didn't make it into the national newspapers because the explanation of her death, when it came, disobeyed too many rules of theatre — causing it to expire, to keep to the theatre analogy, in its off-Broadway run. (The very ordinary Herr Erpenstein, by comparison, had a large audience at his execution.)

Paradoxically, it's the doggedness of the investigator's procedure that gives the story its attractive glow. The shift *away* from the expected outcome throws another light on the poison vessel — not as penetrating as a laser, but certainly brighter than a constable's torch.

Here is what happened to the investigator and how he put the pieces together:

On his first day at the farmhouse he was offered tea and cake. Both produced a tingling sensation on his tongue consistent with arsenic (which he had made it his business to recognise). He emptied the tea into a glass jug for testing, quarantined the teapot and sent a piece of cake to his lab by special post. He questioned the cook, the farmer's wife, and examined the household stores. He told the family to take their meals elsewhere.

Left to himself, the investigator moved across the murder landscape in careful steps. Eventually, no doubt, he would have progressed to the outbuildings and the well, but on this particularly hot day he made straight for the cool shade near the well to write his notes, and there made his first compelling discovery. Looking down he saw the white upturned bellies of two frogs floating on the water surface. The dead frogs are the fulcrum on which the answers balance. The investigator correctly surmised that the frogs had died in poisoned water, and he stood staring into the well feeling, as we all feel, the irresistible pull of subterranean surfaces.

A more romantic (or less pragmatic) man might have widened his search to include farmhands, orchardists, drovers, in fact anyone outside the immediate family who might have got at the water. But there was something in him (something in his life experience, it turns out) that rejected the idea of a secret well-poisoner and held its gaze fixed on the structure of the well itself. Applying his magnifying glass to the bricks at the top he scraped off some samples, then made enquiries about its construction.

The well, his informant told him, had been partly built with recycled hand-made bricks from the goldfields. With this piece of information and the investigator's own experience of mining practices, he put together a definitive geochemical conclusion to the mystery.

Twelve months before the girl died, the farmer had bought a load of recycled bricks to build a new well. He'd had to travel a long way to find them because bricks were in short supply after the war, and he'd ended up at the old goldfields. This was the clue that untied the knot.

In the nineteenth-century gold rush, it was common to find 'fool's gold', an arsenic-containing ore which prospectors burnt off in furnaces, a practice that created a lot of 'arsenic smoke'. Over time the smoke impregnated the bricks. When one of these kilns was dismantled, the bricks were washed and sold. Washing gave the bricks an appearance of cleanliness but trapped inside the porous cavities was a heavy deposit of arsenic. Being only slightly soluble in water, small amounts of dissolved arsenic had been carried into the kitchen for tea and baking, enough to make the family ill from time to time; illness that significantly abated when they went on holiday, or away after harvest.

The young teacher, who liked cups of tea, was a late-comer to a household where the family had been habituating themselves, unwittingly, to increasing amounts of arsenic as the well got lower. The greatest concentration of arsenic was, however, in the bricks at the top of the well, out of harm's way — until two events corresponded. The first was a week of steady rain. The second was the appearance of the seasonal grape-pickers. To get the wax off grapes the workers first sluiced them with potash, or sodium hydroxide, then lowered their buckets into the well to give the crop a final rinse. By a tragic coincidence, the day the schoolteacher lowered *her* bucket into the well was the same day the first grape-washing bucket of the season had been dipped, potash was released into the water and by a 'chemical convulsion' of the potash acting on arsenic, a lethal load of poison slipped out of its cave in the bricks.

This elaborate solution satisfied the authorities that the death could be laid at the door of misadventure. But it did not completely quell the prejudices of the farmer, who continued to suspect his rival.

>–·‹›·–O–‹›··‹

We can take this picture of arsenic as the innocent tool of misadventure a step further by considering its long history as a positive therapeutic choice.

Arsenic as medicine was once a real proposition. Doctors applied clove-scented arsenic paste to skin eruptions. Dentists mixed it with cocaine to destroy nerves in tooth cavities. Asthmatics once smoked arsenic and tobacco pipes to stimulate their bronchial tubes. And a high index of usefulness was achieved when scientists shaped it into a magic bullet for syphilis. (It seems curious that arsenic has had such a tangled and apparently *positive* connection to sex and its sequelae.)

From the sixteenth century until the invention of the

arsenic magic bullet 'Salvarsan', syphilis was treated with mercury pills — a remedy that almost equalled the disease in unpleasant physical effects.

Flaubert picked up syphilis in Egypt, at the end of his travels.[39] Horseback travel caused him some bother, and twice daily he dressed his 'poor prick' until the sores healed. Back home he began a series of mercury treatments that lasted the rest of his life, and greatly affected his appearance and well-being. Thirty years after Flaubert died the spiral-shaped micro-organism that caused his discomfort was found and named.

The journey to the arsenic magic bullet began in 1676 when a Dutch draper, Antonie van Leeuwenhoek, looked down the eyepiece of his new invention, the microscope, and saw 'animalcules' swimming or suspended in different fluids. Some of these interesting creatures were bacteria, but he didn't know that yet. Two hundred years later, Robert Koch, in Germany, not only knew what bacteria were, but also postulated that these bacteria were the cause of infectious diseases. Following quickly on Koch's heels came a new generation of bacteriologists who wanted to know which bacteria caused which disease.

In 1899 a middle-aged German scientist, Paul Ehrlich, moved into a new job in Frankfurt where he finally had the resources to follow through on some of his ideas, his most far-reaching being the hypothesis that drugs could be found that harmed or destroyed the invading bacteria while sparing (relatively) the host. Five years later (after testing over 600 arsenic compounds and finding a winner with number 606), came *Salvarsan*, an arsenic drug that would save a generation from the horrors of the mercury cure (while bringing its own kind of troubles).

We know a lot about venereal disease from authors whose accounts haven't always given a romantic spin to the occupation of their bodies by parasites. In 1914 the Danish writer Isak Dinesen (Karen Blixen) contracted syphilis from her husband, who had a number of Masai mistresses. Her doctor in Nairobi gave her mercury pills to tide her over till she could return to the more modern infirmaries in Denmark. She underwent the arsenic ordeal in a Copenhagen hospital and the disease was arrested in its secondary stage, but it never fully went away. She was thirty years old at the time and would live another forty-seven years. At her death (officially of 'emaciation') she weighed less than 30 kilos, and was still suffering the tortures of spinal syphilis — but had made friends, of a sort, with arsenic along the way, surviving and elegantly describing an overdose that might have killed a novice.

><+>-O-<+>-<

Even though I've read through the arsenic stories and think I've heard it all, I shouldn't be (but am) surprised by the extent of ordinary human stupidity. It infuriates me to read the story of a man who got three women pregnant and persuaded all three to let him put arsenic into their vaginas to 'loosen the child' (each woman died).

But when this sort of stupidity happens in a hospital — in the name of treatment — we have reason to be happy about living in the twenty-first century. The *British Medical Journal* of 1961 reports the case of a healthy twenty-eight-year-old woman who died after her vagina was packed with arsenic pessaries, under general anaesthetic, to treat persistent vaginal discharge. 'In the hospital concerned it has been carried out on numerous occasions without producing any serious untoward effects.' Eighteen Acetarsol pessaries, totalling 72 grains or 4.67 grams of arsenic, were inserted. The following day when the woman became ill she was given chlorpromazine (an anti-nauseant), then after another twenty-four hours *twelve more* pessaries were inserted without anaesthetic on the assumption that it was the anaesthetic that was making her ill. The next day she had a fit, her temperature was elevated, she had a rapid pulse and her mind was confused. The staff washed out her vagina but found it empty of pessaries — they had been absorbed. Half a day later she had another fit, then a few hours later a third. Finally, someone seems to have suspected poisoning and administered an antidote, too late. Several more fits deepened into a coma then she died.

>—I—◆>—O—◆>—I—◅

Modern sightings of arsenic are mostly confined to industry or working specialties licensed to carry poisons. In the 1980s a St Louis man who worked as a hospital lab technician swallowed arsenic (which we know works slowly and painfully) then injected himself intravenously with a barbiturate. The sedative sent him into deep unconsciousness while arsenic did its dirty work at leisure.

The old queen has recently made an appearance in folk remedies from Northern Laos, discovered when immigrants were hospitalised in the United States and Canada. And she's been found in ethnic pills in one of her old principalities, India.

But the saddest final glimpse of arsenic abroad was caught on the back streets of Miami. An eighteen-year-old male cocaine abuser was admitted to hospital (after a crack binge) in a very poor state. His hands and feet were numb, he couldn't feel painful stimuli, his deep tendon reflexes were absent. Unexpectedly, the tox-screen showed arsenic at twelve times the normal level. Hair samples showed regular exposure to the poison. Pleased with their detective work, the doctors notified their patient of the extra danger of a poisonous additive circulating in street cocaine, only to be met with indifference, and the news that it was common knowledge that arsenic was used for cutting.[40]

><+>-0-<+<

In my bottle collection I have several arsenic specimens, a little cyanide bottle, and a neat little tin half as high as a tea cup, labelled *Strychnine*. I keep them in my dining room, in a locked cabinet — because it's common sense, and because rules for storage, as laid down by the Poisons Act, were drilled into me at work. *Never store poisons near food.*

But I do. When friends come to dinner we sit at a table only a few feet from my deadly souvenirs. The conjunction of these two collections — my friends and my poisons — has no menacing overtones for me; we could be sitting in the garden, near the shed and its containers of weedkillers. We eat, we talk, the bottles sit on their shelves and the world turns.

Lately, however, a thought bothers me and I find myself looking over my friends' shoulders at the shapes behind the glass. Could I be unconsciously replicating

Agatha Christie's Mr P with his lump of curare? Is my collection an unconscious expression of some sort of attraction to *power*? Is this the slippery rock where my grandfather lost his footing?

Chapter Seventeen

The demon nurse

AS A WOMAN, I CAN'T help noticing how often we're portrayed as the cunning (and peculiarly unique) exponents of poison craft. At some level, everyone — except perhaps the odd professor of criminal justice who's studied statistics — seems to believe the generalisations about our kind: that poisoning is women's work, that poison is a woman's weapon and that women haven't got the *cojones* to kill like men. Literature, science and popular entertainment all endorse this image of 'dangerous woman'. The wicked witch, demon nurse or whatever you care to call her persists in the cultural landscape like a radioactive isotope (an entity from a scientific discipline which, coincidentally, identifies the products of radioactive decay as 'daughter nuclides').

Like a fascinated eavesdropper, I've deliberately tuned in to some of these frequencies. The news is categorical, unambiguous. Women can't be trusted. Our dishonesty hides its face as we turn to fasten our apron strings; our double-dealing triumphs when the sick gratefully open their mouths to our spoons of medicine. What position can an innocent representative of the gender take in the face of damning evidence? — the pride stance? *yes, we're good at it so watch that cup of cocoa*; the outraged feminist defence? *get off our backs.* I've no idea. And I'm not entirely innocent of poisonous thinking myself.

>—◆>—O—◆—<

As both woman and chemist, I'm doubly cursed. In *The Criminality of Women*, Otto Pollack writes, 'with woman's social roles as the preparer of meals and the nurse of the sick, it appears obvious why poison should have become the favourite weapon of female murderers... physicians and druggists resort equally to poison.'[41]

Pollack posits three sexual summits of a woman's life — the menarche, pregnancy and menopause — that correspond *statistically* to peaks in criminal activity. In the first and last peaks, homicides (by methods including poisoning) are due to 'irritability', while the middle period promotes 'attacks against the life of the foetus and the newborn'. He concludes his book with the sentence, 'In short, the criminality of women reflects their biological nature in a given cultural setting.'

This extraordinary picture of the inherent toxicity of women never quite goes away. It started long before the first Christian woman, Eve, and it's been downhill ever since. Kipling generalised across species (the female is more deadly than the male); Nietzsche said women were 'barbarous' in love and revenge (the most common motives for poisoning); Professor Glaister (writing in general terms about real cases) claimed that women have always set a high standard in novel methods of poisoning. Chesterton felt that poisoning fell naturally among the domestic duties of women;[42] Machiavelli, who excluded women altogether from his 'art of war', nevertheless put a grid across the danger divide between men and women when he wrote about what is now called by psychology 'the strong visual sense of the male'. A man going into battle (as opposed to the devious dangerous woman) changes his way of dressing, his voice and his customs. He discards civilian effeminacy, in the cause of 'terrify[ing] other men with his beard and his curses'.

By creating both believably unbeautiful witches and beautifully evil women, Shakespeare has set a context for real discussion. If she looks like a witch, smells like one and acts like one, we can reasonably expect witchy behaviour. Shakespeare's witches don't let us (or Machiavelli) down in that department. Skinny-lipped, bearded, the weird sisters in *Macbeth* prance about the stage with all the bells and whistles sounding. There's thunder, lightning, invisible familiars, curses and predictions. Witch number one threatens:

> '. . . like a rat without a tail,
> I'll do, I'll do and I'll do.'

And she does. For a study of the evil in a titled, elegantly dressed woman one only has to hear Lady Macbeth speak the line 'Leave all the rest to me.'

By comparison, the deadly female who *doesn't* dress the part, who in fact looks and smells like your mother or daughter, becomes the truly terrifying spectre, the blank space into which we project our fears and onto whose unremarkable physiognomies we feel compelled to mentally sketch the missing elements that will turn her into a witch.

Once you start looking for these dangerous women the material unravels by the yard. Literature and art love them. The internet loves them. I found a site that lists all the people known to have ever been accused of witch-craft and 'killed because of it' — 236,870 in total. They're not all women but most are; the men who arrived at the stake came on women's broomsticks.

I come to these women to see what we have in common rather than what differentiates us, a bit like an anthropologist who chooses to spend time with a tribe who — superficially at least — look and sound like her first cousins. My first choice was someone modern, a witch of our times, and I wanted the luxury of dissect-ing my specimen with my own tools. To achieve this, I went to the files and simply looked at the artefacts. I listened to no one else's opinions and sought no other interpretation.

My demoness was Simone Weber, born 1930, senten-ced to twenty years imprisonment in 1991.

I looked at two pictures in a magazine layout, posi-tioned so that as you turned the page you saw first a full-page shot of Simone Weber, captioned 'the devil in the dock', then on the right-hand side of the facing page a smaller shot of a crowd waiting outside the courtroom in Nancy, France. This caption said, in part, 'Crowds trying to catch a glimpse of the She-Devil'. Side by side, these images are shocking, and subverting.

Madame Weber, accused of two murders, is an ordinary, dumpy matron with permed hair. Her eyes, turned towards the photographer, show what might be anger or surprise (or heartburn). Two policemen, one behind and one beside, crowd her into the cramped space of the dock.

Moving to the words, I read that the prosecutor said, in her opening address, 'It is enough to look at you to understand that if the devil exists, you have sold him your soul.' My gaze flicked back to Simone, this time looking for what I'd missed. I narrowed my eyes and searched, hard. White blouse, white cardigan. Nothing in the high forehead, or the large ears and double chin. What about the eyes? — ('Exophthalmia' was one word that came to mind.)

The second photograph, the crowd scene, is taken from a first-floor window, at least six metres above the road. In the bottom right-hand corner of the picture is a doorway against which the first two people in the queue are pressed. The line goes off at a forty-five degree angle to the door and there are six people leaning against a barrier. Behind the six and forming a wedge-shape is the first flank of the crowd, who have blown out the single or double file intention of the officials to a shape of their own making, following some crowd-dictated law of entropy. On the far left are six obvious push-ins, three couples who are off the footpath and on the road and determined to beat the obedient queuers who have kept to the footpath. There's snow on the ground and everyone is thickly padded. What's disturbing in this picture, much more disturbing than Simone Weber in the dock, is the beaming smiles of the group in the front wedge. World Cup grand final ticket holders couldn't look more pleased with themselves. The sense of antici-pated pleasure that unites and holds these people in very close contact on a footpath in the freezing cold for an unspecified length of time is as fresh and strong (and discomposing) today for the observer as it must have been in 1991.

Studying the faces with a magnifying glass I see a similarity of expression many of us have witnessed at public spectacles. Or read about and imagined. Tacitus paints a picture in *The Burning of Rome*: 'On his way to the Adriatic, Nero stopped for a while at Beneventum. There large crowds were attending a gladiatorial display given by a certain Vatinius.' The gladiator, a deformed and witless 'monstrosity', was a great crowd pleaser. Between the lines of Tacitus's deadpan account you can imagine the spectators straining to see the gargoyle head of the gladiator who had risen from being the butt of jokes to a position of influence and wealth.

Simone Weber's crowd reminded me of Robert Capa's famous photograph of a French woman collaborator, head shaved, holding a baby, hurriedly walking down a brick street, trailing a party of excited townspeople. Capa's image, chillingly composed, kinetic and emotionally charged, is superior in every way to the crowd scene in Nancy, yet the connection between both groups is not hard to make. Capa's crowd is on the move, you can hear them puffing to keep up. The trio of women to the left of the shamed woman are almost gleeful. The front-runner, a girl in bobby-socks, looks as if she's just drawn level with Frank Sinatra after a desperate pursuit. At least eight of the people closest to the collaborator have ear-to-ear grins.

Simone Weber was said to have poisoned her husband with digitalis. A year later she began a relationship with another man, who, it was alleged, she tried to poison (because he was unfaithful) and who disappeared suddenly one day in 1985. The evidence against Simone suggested she had sedated her lover, then shot and dismembered him with a chainsaw. At trial she was acquitted of poisoning her husband, but got twenty years for the murder of her lover. As her counsel put it, they had a poisoning without a poison (toxicology couldn't find a trace of digitalis) and a murder without a body (a torso, found in a weighted suitcase in a river, could not be definitely identified as the lover).

Before closing the file I looked at a shot of the family of one of the dead men. They are sitting in attitudes of grim accusal, arms folded, one woman with her head sunk in her hands. How much easier it must be, I thought, to have 'right' on your side — to paint a witch when you've got a thick palette knife and pots of coagulated paint scavenged from the crime scene.

Modern non-witches who look like grandmothers have, by and large, had an easier time of procuring their poisons. Chemist shops, doctors' surgeries, hardware and even gardening centres provide handy pick-up points for the modern poisoner's needs (the most up-to-date buy theirs off the internet). Simone Weber used stolen prescriptions to obtain digitalis, with which (it was alleged) she over-dosed her husband of three weeks (an eighty-year-old who had advertised for a companion in a lonely hearts club), and later gave it to her lover who began to have dizzy spells after things soured between them. Digitalis — the foxglove drug — doesn't leave lasting evidence like arsenic. The drug is degraded and excreted from the living body within six weeks, taking with it all traces required for identification by a forensic investigator.

What do we imagine these poisonous women to be like that makes the real thing so sordidly unsatisfying? We all know what a witch looks like and what she does, even the least dexterous of us could draw one with a pencil. But seeing the woman in the witch is a more difficult exercise, achieved most successfully, and paradoxically — it seems to me — by novelists and poets.

American author Nathaniel Hawthorne has given us a template of sorts in his short story 'Rappaccini's Daughter'. Young, virginal, cloistered (the way Gothic girls are), Beatrice is the human sister of the poisonous flowers her scientist father grows. 'A being of classic fable who lives on sweet odors', the girl exhales a poison breath, and like the unnatural plant she is twinned with (the result of her father's 'fatal science'), she is as terrible as she is beautiful.

A gift of fresh flowers withers in her grasp. The touch of her hand leaves a purple imprint on her would-be lover's wrist. This lover, Giovanni, — young, blond and ardent (the way Gothic boys are) — is both fascinated and repelled by Beatrice. His feelings are not love, not horror, 'but a wild offspring of both'; and, inevitably, he is drawn into, and infected by, the forbidden atmosphere around Beatrice and the plant with flowers like purple gems, her 'sister'.

By getting too close Giovanni is stung. This is the nub of our human relationship to what is perilous: if we go too near, we get hurt — yet knowledge, reason and all the weapons of our self-protective rational minds won't stop our fingers from itching to stroke the purple gems.

After he is infected, the young man begins to manifest the same poisonousness as Beatrice. He breathes on a spider and it curls up and dies. He buys fresh flowers and they wilt in his hands.

Turning to a senior medical man at the university for help and guidance, Giovanni unwittingly plays into the hands of Rappaccini's rival and detractor, Professor Baglioni.

'Behold this little silver vase,' says Baglioni, handing Giovanni a phial of powerful antidote 'distilled from blessed herbs'. The gift is represented as a means of delivering the couple from the grip of Rappaccini's experiment, but is in fact Baglioni's revenge on Rappaccini's insane zeal for science.

Beatrice swallows the antidote, and because she is poison herself, *she* is annulled, and drops at their feet.

The seductive attractiveness of Beatrice works at many levels. My own obsession with poison owes some of its depth to the high magic of Beatrice's kind of witchery. A pretty girl, a locked garden, a strange and deadly plant: these are hooks that snare imagination and hold it in thrall.

And there is Beatrice's warning to Giovanni, her lover, when he gets too close to the poison plant: 'Touch it not. Not for thy life! It is fatal', a variation of the words my father said to me the day he opened William's travelling case and I got too near my grandfather's bottles.

It's the Beatrice-like qualities I'm searching for when I peer at photographs of my poison women; like Giovanni I want to experience a horrified love at what I see. And invariably I am disappointed.

As a counter-blast to the dull thud of Madame Weber, I turned to the seventeenth century to examine three 'authentic' witches: Marie Bosse, Catherine Deshayes, and Marie Vigoreaux, all women who poisoned for profit.

'Only three more poisonings and I can retire!' Marie Bosse is supposed to have said in 1669. These women not only dressed the part and kept 'laboratories' for making potions, they indulged in black magic with so-called obscene priests, and practised incest and ritual sacrifice (or so it was said). They turned abortion of royal bastards into a profitable sideline. When Deshayes' garden was dug up, the remains of multiple unwanted 'aristocratic' children were found. Because the king's mistress, Madame de Montespan, was implicated, a closed court was assembled to try the case, and over thirty people put to death, the same number banished and four sentenced to life imprisonment.(The aristocracy who largely financed this backyard trade naturally escaped public consequences.) Marie Vigoreaux died under torture. Bosse and Deshayes were burned at the stake.

Witch burning was the ultimate public spectacle. The seventeenth-century mania for scrubbing out all traces of a superstitious past, fed by the clergy and newly created faculties of medicine, gathered up ordinary herbalists, midwives, lay medical healers, accused them of dark deeds and, every so often, burnt one or two or three. Whether the punishments fitted the crimes is still being debated; what is certain is how much 'easier' it was to shout *Burn the witch!* when the woman concerned was considerate enough to act the part.

Looking for an Eastern picture of the Western devil woman, I found the Indian legend 'Krishna and the Demon Nurse'. The seeming self-contradiction of the word 'demon' nestled up to the word 'nurse' creates the effect of untrustworthy nurture that throws its shadow across the poisoning crime, and makes it so *feminine*. The story goes that Krishna, a dynamic incarnation of

Lord Vishnu, was wet-nursed by a demoness named Putana (or Pootana) disguised as a gentlewoman who'd lost her children and husband in a famine. Putana painted her nipples (or filled her breasts) with poison and put Krishna's mouth to it, confident that the six-month-old child would soon die in her arms. However, Krishna — being no ordinary baby — sucked with such supernatural vigour that Putana got the shock of her life. She tried to push the baby off her nipple but he gripped even harder, sucking out her energy until she died, and in death revealed herself in her true, pimply (and smelly) demonic form. Much is made of the unmasking of a disguised witch before any damage is done in this legend. Clearly, in the Eastern version, when a demoness meets her match, outcomes may alter.

Breasts, of course, are as much a part of the poison picture as cooking pots. In a 1988 case cited in the literature, a sick eleven-day-old baby was brought into a public hospital. The child was blue, tachycardic and after admission experienced major seizures that no amount of sedatives could control. The urine tox-screen showed a high level of cocaine, a substance the mother denied using in the accepted sense of snorting or injecting, but later confessed to using as a dusting powder on her nipples to relieve the pain of breast-feeding without washing it off before suckling, adapting cocaine's well-known topical anaesthetic properties with success to herself and almost fatal consequences for her child.

Poison women of later centuries — notably the nineteenth and early twentieth — are mostly to be found at the sleazy end of town, not, unfortunately, in brocaded halls, or laboratories for necromancy, or nannying in an avatar's nursery. (Although the odd carriage may clatter past bearing a pretty lady in an arsenic dress on her way to a ball.)

Poisoning hit another high note in these times mainly due to the easy availability and cheap price of arsenic. Sad and desperate wives took on the mantles of witches with unimaginative expediency, or they simply wandered off the domestic path into the fatal domain of murder.

Like illiterate, immoderate Charlotte Bryant, who was hanged in Exeter prison in 1936 for feeding her husband an arsenic-laced Oxo drink; or Ethel Major, who spiced the corned beef with strychnine and was hanged in 1934. So many sorry, reckless stories, so little inherent evil.

Returning to the files hoping for a story with a bit more *oomph* than a defeated wife at the end of her tether, I found Mary Ann Cotton.

Mary Ann Cotton was a poisoner for profit, who might have exclaimed with Marie Bosse, A few more murders and I can retire! Mary poisoned three husbands, her own first eleven children and four others in her care, a lodger, and (probably though not proven) her mother, who died in 1866 shortly after one of Mary's visits. All the deceased were insured for a few shillings with British and Prudential. Her motive was greed, her method arsenic, her opportunity the porridge pot.

Mary embodies the nineteenth-century version of the toxically dysfunctional demoness hypothesised by Pollack, but what a disappointment she is to look at. Drab in her cloak and bonnet tied under the chin, she stares open-mouthed and hunch-shouldered at the photographer (that new intrusive witness to shame). Where, one wonders, is the visible clue to so much heartless behaviour? Policemen, judges, journalists and historians all tried to get a handle on Mary. Much was made of her childhood of chronic deprivation, her father's early death (he fell down a mine shaft) and her mother's hasty remarriage. Mary married four times, always with an eye to money and social improvement. She advanced from wife, to mother, to employed nurse, to wealthy man's housekeeper and finally to landlady, each one a step away from what was for her, given her class and the times, preordained poverty.

At some point during, or before, her first marriage Mary acquired a self-taught connoisseur's appreciation of arsenic. It was all too easy for her. A term or two of nursing at the Sunderland infirmary provided a ready-made glossary of convincing causes of death: Mary's husbands died of 'gastric fever'; her babies of 'teething convulsions'; the older children of 'unspecified fever'. Mary took care of it all: spooning out the arsenic, nursing the sick ones to death, collecting the money (in inverse relationship to those around her who — like the wise monkeys — saw, heard and said nothing).

And, like most compulsives, she overstepped the limits. Wanting to marry again after disposing of the last husband, she tried to place her surplus-to-needs seven-year-old step-child in the workhouse. When the workhouse refused to take the boy, Mary gave him a lethal dose of arsenic and called for the doctor to verify death from gastric fever. There was nothing untoward or sloppy in her technique with the boy's death, it went as all the others had, with cries and pain, and impatient waiting for the breathing to stop, and yet, as if this was just one too many affront to the gods, a neighbour acted on his suspicions and called the police. There were inquests, bungled autopsies, exhumations, a chemist who remembered selling her twopennyworth of arsenic for 'bedbugs', and finally, as a kind of dramatic relief, the trial, and imprisonment.

In an ironical postscript, Mary gave birth to her twelfth child, a girl, while waiting to hang in Durham gaol. Even this step, the final one that would take her to infamy, was accompanied by the inconvenient birth of a child. She was hanged in 1872. Nobody shed a tear.

If the dancing lady in the arsenic dress is one of poison's martyrs, Mary is one of its pathetic despots, not quite in the league of her medieval predecessors, but sordid enough to take her place in the procession of toxic women.

Prettier, richer, and better connected was another Mary, a French magistrate's daughter who married a minor nobleman and became a Marquise. Marie-Madeleine d'Aubray (the Marquise of Brinvilliers) fed her father arsenic-laced soup over some months until he died and she could realise her part of the inheritance. Like Cleopatra, Marie approached poisoning with intelligent and cold-blooded experimentation. Her lover had been detained in the Bastille, where his cell-mate revealed himself to be (conveniently) an Italian skilled in toxicology. When the lover was freed he teamed up with a chemist and set up a sort of laboratory, where Marie collected samples and took them on her rounds of charity hospitals, practising her own form of *noblesse oblige* by trying out doses and watching for effects on the charity patients (called 'the poor and the sick', as they no doubt were, by narrators).

Taking quite a shine to the art, she then poisoned her two brothers and collected their shares of the inheritance, enhancing arsenic's reputation on the streets as 'inheritance powder' (*poudre de succession*). In 1676 Marie was put on trial for the murders of her father, brothers and husband, and an uncounted but significant number of charity patients. She was tortured on the rack then beheaded. The Inquisitor had a lot of fun with the Marquise, as did her torturers. There are two drawings of Mme Brinvilliers on the rack — naked in one; in the other, decently covered by a nightgown. In the first — undertaken, one can't help thinking, with a degree of sado-masochistic projection — the nude Marie is on her back, her head and feet drawn down, projecting forward her breasts and pelvis for the entertainment of the quill-holding cleric/priest whose gaze is aimed directly at the Marquise's pudenda. The torturer pours water into a funnel in her mouth.

In a 1996 study of nearly 300 homicides by American females, there wasn't a single case of poisoning.[43] Women still killed but in this study they used the so-called 'men's' weapons: guns and knives (how does this relate to the biological determinism argument?). But the poisoning 'type' never really fades away; this under-reported crime endures like arsenic in churchyard soil.

Velma Barfield was executed by lethal injection in North Carolina in November 1984 for multiple poisonings. She once had the distinction of being 'the last woman executed in the United States', but enjoyed none of the other claims to fame of her poisonous sisters. Velma was fifty-one and a grandmother, not photogenic, not a TV face like Karla Faye Tucker (who dethroned Velma by being executed in 1998), and not a patch on Beatrice Rappaccini.

Further away still (though approaching the witches in Act I Scene i of *Macbeth*) is Elfriede Blauensteiner, tried in 1997 for poisoning twelve rich old men with a diabetes drug, Euglucon. Elfriede, nicknamed 'The Black Widow' and described as 'the most venomous female poisoner in Austria since the Middle Ages', has a female fan club who applaud Austrian women standing up to their men, by whatever means.

The poisons of the late 1990s are prescription drugs; there are no more arcane laboratories, no secret gardens, no purple gems. The new laboratories make speed from pseudoephedrine, crack from cocaine. The new botany hybridises marihuana.

An Aboriginal woman once asked me, 'What is your poison dreaming?' I was stumped for an answer. The Northern Hemisphere, I supposed; Greece, Rome, the Renaissance, the nineteenth century, Hollywood, Disney? I could only think, at that moment, in terms of geography and fashion. She might have been prodding me to think about other kinds of knowing, the sort it's possible to receive from invisible patterns and energies that have been here since long before the earth was formed, and which are brought into consciousness by stylised storytelling.

I looked at a lot of Australian Dreamtime stories and found one that tells of two *wirreenuns*, or sorcerers, mother and son.[44] There is no poison in this story, no stinging plant or snake or rock, to identify it with the familiar profile of the venomous woman. It's a story that's repulsive in its literal rendering. A mother who is entrusted with caring for her son's sick wife tricks her into falling asleep beside their baking dinner, then bites the girl's nose off and leaves her to bleed to death. The son, who is secretly a greater sorcerer than his mother, discovers the tragedy too late to save his wife. In great sorrow he sets fire to the spirit tree, burning his mother alive, then goes off 'on his solitary way'.

The brutality of the son is no less shocking than his mother's, and in a perverse way reminds me of the unmasking of Putana by Krishna's vacuuming lips. In an elegant interpretation by Johanna Lambert, I learned that the issue at the heart of the story is a mother–son relationship that has outlived its biological shelf-life, an unnatural place where a grown man hangs on to his mother's apron strings when he has a young pregnant wife and is of an age to be getting on with a new independent life. The themes are male initiation, severance and the devouring mother, the feminine that destroys in order that a cycle may finish and a necessary new thing can grow from the remains of the dead; what has been called in one context the 'female power' to deceive and destroy man, and was called by Freud the Oedipus Complex.

In these devouring figures, like Kali and the *wirreenun* woman, are the patterns appropriated by psychology as the negative aspect of the mother archetype, the poisonous woman.

>-+-◦-+-<

I still dispense, in the mixing-things-together sense of the word. My fingers are comfortable moving to learned rhythms with a spatula. There is a persevering way of wetting and crushing sulphur powder until it yields, with a hell-stink, to the soft, identity-annulling corpus of a cream base; there's a sort of dignity in the way menthol crystals mash into snow under the pressure of a pestle and release the aroma of an alpine forest. Pressing the spatula back and forth, *incorporating* the lumps, is something like the blending stage of cooking and yet not. The delineation between dispensary and kitchen is always clear for me; it has something to do with temperature — the kitchen is the warmth of the hearth; the dispensary is the clean cool of the medicine chest.

In my roles as woman and chemist I've never knowingly poisoned anybody, but I've thought about it.

A woman whose face I can barely bring to mind now momentarily woke up the sleeping poisoner in me one ordinary spring day. She used to come into the pharmacy in all seasons and levels of humidity, insisting that her bottle of special cough mixture be 'made up'.

The mixture contained a high level of codeine to which she was addicted, and like many addicts she was overly sensitive to any hint of non-supply. Counselling her was like stepping on a cobra.

On this day, in the middle of assembling ingredients — codeine syrup, a flavouring balsam, a dash of red colouring — I moved some things aside and noticed a ribbed brown bottle tucked in at the back with its face to the wall. Turning it around, I read *Arsenic solution*. The moment seemed to stall, like a held breath. As time expanded and lengthened, I deliberately held the bottle over the measuring cylinder and wondered what a drop or two would do, how much would hurt, how much would kill, how I would cover my tracks.

Ultimately, and naturally, I did nothing remotely like adding arsenic to her cough mixture. Not being a psychopath or a sociopath, I would need the motivating energy of a burning grievance, a deeply felt passionate emotional impulse to kill — certainly something stronger than simple annoyance. What the moment gave me, I now realise, is a glimpse of the terrain, the mental landscape in which I *might* add the 'vile pinch', minus the clinching factor, an emotion as strong as hate or love. To steal a line from Kundera's *Immortality* (and change the pronoun), 'I cannot hate her because nothing binds me to her; I have nothing in common with her.'

Chapter Eighteen

The ghost in the labyrinth

I'VE ALWAYS HAD THE sense that what puts us in the path of danger, as well as what separates us, is the same thing: a small volitional act, as flamboyant as smashing a glass, or as subtle as brushing away a mosquito with virus-loaded blood.

For the ordinary person who might think about poison once in a while — say, after a news report — the old mysteries and doubts we all share about how we will die usually make a brief but memorable appearance then fade away. It won't happen to us. Statistically, we tell ourselves, there's more chance of being crushed by a meteor. Anyway, we'd know, we'd guess, we're wiser than the easy targets of past centuries.

The trouble is, not being paranoid by nature or inducement, the majority of us won't suspect a thing. The thousand acts of faith that underpin what we do in our day — eat food we didn't prepare; eat food we prepared but didn't grow; lick our fingers; garden with chemicals; wait on corners breathing fumes; mis-read medicine labels; swim in rivers; accept liqueur chocolates from friends at work — are also small acts of will.

Poison of itself is no bad thing. A toxic substance is quite happy minding its own business; it has no designs on our bodies and no capacity for conscious thought. It takes the action of a vector to bring poison into contact with flesh, and a vector can be any unreliable go-between, any person or thing that leaves the gate to the labyrinth open.

Once the minotaur is loose, we remember the gate.

Consider this small story from England.

A man tires of his wife. She has a lingering illness, is often in hospital, ill and ailing but stubbornly alive. The man takes a lover. He begins bringing his wife a gift of poisoned cake, and after each visit the wife becomes sicker. In the context of her illness and accepted notions of mortality, her deterioration is noted and accepted, and when she finally dies there is a gratified sense of ordinary conclusion.

But even in the twilight zone of morphine-induced stupor, the wife reads the changed language of her husband's body. Betrayal betrays itself.

With the last of her strength she writes a letter to a police constable and encloses a piece of cake, in a box tied with string, for him to analyse. The note and box become separated. The constable opens the box, sees the cake and eats it. This is the volitional act, the step that positions flesh for its dose of poison. (The letter *does* arrive the next day; and the constable survives, leaving us a graphic account of the worst night of his life.)

We can understand the wife feeling flattered by her husband's gifts and showing good faith, both to him and her own fading body weight, by eating it in his presence. But what are we to think when she takes the second and third piece and, after each, spends a night of intolerable pain that is diagnosed as the progression of her illness but which she *knows* is the criminal ministering of the man who's supposed to love and protect her?

><+·+»·0·+«·+·<

Most examples of what we call *poison* today are inanimate powders and liquids. Left alone, they persist in some molecularly stable way, their internal bonds as intricate as chain-mail. Pure arsenic for instance is an element of the earth domain, bound in varying ways to other earth elements, inhabiting its own inorganic cosmology. When a volitional act spins arsenic into another galaxy it is treated like an invader and dealt with accordingly. The human body scans arsenic as a potential food source. It unshackles arsenic's bonds and attempts to burn the parts for sustenance.

With strychnine, the switch to a nerve pathway is too quick for the alimentary engines. While they are getting ready to fire up, the main switchboard in the brain is already shutting down. Strychnine and flesh make a hectic, fatal pairing. The minotaur wins with one shattering pass.

Whose act, and with what degree of volition, put strychnine into Thomas and Patrick's mouths? This is the question behind all my questions, and the story ghosting all the stories in this search.

Chapter Nineteen

A supply of white rats

A DISTURBING CATALOGUE of animal and inverte-brate experimentation underscores most of what is known about toxic substances. I can't pretend to too much squeamishness about it, having been a student myself at a time when one tested the hypothesis that morphine dulled the perception of pain by injecting mice with the drug and observing their feet quietly smoulder on a laboratory hot-plate. It's impossible to work in any field of pharmacological research without a breastplate of clinical detachment; and those who lose their shields wisely step out of the arena.

Before 1900, experimenters dropped poison into water-filled beakers where they could watch the effect on tiny aquatic protozoans. While it was apparently fascinating to see them swell and burst, nothing of value was gained because the swelling and bursting didn't approximate human reactions.

More complex aquatic organisms — cuttlefish, squid, octopus — reacted obligingly to doses of strychnine:

dead in two minutes, corpses rigid, tentacles stiff, hearts still beating. But, again, extrapolation to human reactions remained limited by the very different physiology of the subjects.

Dr Blyth was particularly fond of blowflies. Their articulated legs told him a lot about gait, but nothing about the brain (blowflies have no eye pupils) and less about heartbeat. Larger, more multiplex organisms were required, and the dissectors turned to frogs, which are efficient at showing changes in respiration and nervous conductivity, and have the convenience of foot webbing for viewing blood circulation in the periphery.

Inevitably, however, mammals had to be strapped to the bench if the mysteries of mammalian physiology were to be elucidated.

Out of the horrors of this early mammalian experimental era (I'm thinking of a torture-chamber device called 'Czermac's rabbit holder') came the white laboratory rat, and the beginning of what Nabokov (in *Lolita*) called the 'middle-class nosy era' of the 1950s where 'you have to be a scientist if you want to be a killer.'

The 1950s and 1960s are significant in the history of poisoning. Although this period marks no *end* of the practice (poisons and people are still with us) it represents a hurdle that made the game no longer worth playing for the ordinary performer. In one corner scientists were busy inventing identity tests and antidotes for poisons, in the other corner, societies were legislating against easy access. Many, like Nabokov, sighed with regret that it was no longer possible to drop into your local chemist, sign the register, and leave with your packet of poison. Most dispensaries still had a back shelf, just above eye level, where you could find arsenical solution, strychnine, strong acids, and murderous looking tinctures, but they were kept at bay under the new, 'nosy' requirements of the Poisons Act.

In an old seaside pharmacy I once came across a 1950s Poison Register. The pharmacist of the day had signed out half-ounces and ounces of strychnine, arsenic, cyanide, thallium and mercury to the local citizenry. In all but one case the poisons were used on animals — domestic and wild cats, kittens, rats, birds, rabbits, and dogs — and 'pests' like white ants and borers. Meditating on this carnage, I read the single eye-catching exception: 'W. Parsons, Fishmonger, Ettalong, half an ounce of arsenic trioxide *for experimental use.*'

In the same archive I came across a 1953 letter from the headmaster of a public school requesting the chemist to make up a 'Killing Bottle', a chamber for killing insects that allowed for 'quick mounting while the insect's body is still pliant', and reminded me of other chambers where a quick kill was desirable. I could find no evidence that the chemist complied with the request, and guessed that he would have felt hesitant about handling such an intensely poisonous substance, although the British Pharmacopoeia records that entomologists regularly used this approved method of mixing plaster of Paris cream and potassium cyanide to create a bottle that is constantly filled with poisonous vapour.

After 1966 in Australia, the buying of poison by members of the public, for any reason, was prohibited by law. If by some subterfuge, or special dispensation, you obtained your bag of powder and managed to serve it up to your victim, the intervention of forensic snoopers soon uncovered your tracks. The poison stage was now dominated by a 'crime chemist' with a sophisticated armoury of machines and a supply of white rats.

In the school of pharmacy where William Macbeth learnt his chemist's art, he may have seen some of these early examples of 'life experiments', and, like me cooking mice on a hot-plate, he might have been more affected by the elation of working in an inner sanctum than by any private sensibilities. The excuse, I expect, is youth. Ford Madox Ford (in *The Good Soldier*) might have been speaking for us both: 'twenty-four is not a very advanced age. So she did things with a youthful vigour that she would, very likely, have made more merciful, if she had known more about life.'

Chapter Twenty

Their unflinching gaze

I HAVE A HEALTHY FEAR of poisons. When my ten-year-old self held the stopper of my grandfather's bottle to my nose and was warned *Never, never do that, you could die*, a powerful codifier lodged in my brain. It's still there, a neural mechanism implanted early and kept in order like an alarm circuit waiting to ring its bells.

In my case this fear hasn't converted into an anxiety disorder and I don't know who to thank for that; it might be genes or conditioning or luck. We know that ordinary fears *can* become extraordinary, that a normal fear circuit can go into overdrive and stay in the red zone whenever a stimulus sets it off, and that sufferers need major interventions to re-route these exhausting excursions away from hyperactivity into everyday blips on the screen.

A friend's daughter (whom I will call Eve) ate a spoiled hamburger one day and over the next five years evolved a complicated, ritualistic approach to eating, coupled with an irrationality over hygiene that turned her into an emaciated recluse. One uncomfortable night in a hospital, one or two injections — somehow these reasonably commonplace events detonated a neural explosion in Eve's brain which in turn created new hard-wiring sensitive to the food stimulus. Forever sensitive, it seems. Years of therapy and medication have improved her outlook, but the beast is only tamed. Apparently the rogue circuitry can never really be erased.

The rituals Eve devised (separating her cutlery from the group cutlery; washing and handling plates in special, secretive ways; wearing gloves, and so on) became part of her internal reality, the etiquette of her private table. In the context of the history of poisoning, she is performing tasks which were once externalised and imposed on a separate group, the poison-tasters of the great houses of Europe. Eve's idiosyncratic protocols — quarantining utensils, insisting that foods not touch each other on the plate, protracted hand-washing — might have been relegated to another to perform on her behalf if she'd lived in a noble house in medieval Florence. What we see now as neurotic and unnecessarily self-protective about Eve's attitude to food was, only a few hundred years ago, a highly formalised and choreographed overture to any meal served at high table. The poison-taster acted like an assayer. He might touch the food with a crystal that changed colour in the presence of poison; he might sniff the meat or taste the gravy or scrutinise the wine for sediment; and none of these graceful activities carried out in full view of the seated guests would have raised an eyebrow.

The fear of being poisoned is called 'toxicophobia'; the fear of poison itself 'iophobia'. *Ios* (a spear) and *ios* (poison) meet in sense and pronunciation in the work of jungle arrow-makers. Much of what we read, and indeed are thrilled to read, on the subject of poison works through the agency of fear: the idea of a chemical or plant terrorist secretly invading our bodies, breaching our defences before we have any idea the gate was left open. The difference between the two fears — fear of being poisoned and fear of poison itself — is not a hair-splitting distinction, but a description of the paranoid minefields some of us, like Eve, wander into.

⋗⋯⊙⋯⋖

When Ellen Macbeth left her husband to live with her sister Rose, she too strayed into a domain that fostered suspicious thinking. Now reduced to living on her sister's charity while William played the high life to the hilt, Ellen seems to have gone into a sort of physical hiberna-tion, reverting through exhaustion to the younger sibling role, letting Rose present the face of outraged woman-hood to the world. My father was old enough to remember some incidents from these years with Rose. What he describes fits the pattern of a developing toxi-cophobia, directed against William and instilled in the brothers as a fear that their father would try to poison them if he got the chance.

My father knew William in life only briefly as the man who turned up at Rose's house at odd times and caused his mother to speak in a low, tight voice. On his rare visits William took his sons out for a drive in his expensive and much noticed (by the neighbours) car. The chauffeur kept his eye on the boys (who leapt about the cabin opening and shutting the curious compartments and twisting anything twistable) while William sat benignly in the back, smoking. When the outing was over Rose would haul the boys inside. Had their father given them anything to eat? Drink? Had they obeyed her orders to accept only presents of cash money?

My father remembers Ellen lying on her bed in the dark during these scenes.

>─┤─◆>─0─<◆─├─<

Ellen emerged from her dormancy gradually, and after some time, perhaps a year, William stopped coming around. I found out why when I looked through the newspaper archives at the State Library. In the *Sydney Morning Herald* of 1931 I found this notification:

> An Application made under the Married Woman's Property Act by William Macbeth, herbalist, of Bulkara Road, Bellevue Hill, for a declaration of ownership of a motor car, motor cycle and other property held by his wife, Ellen Macbeth. The wife alleged that the husband had given her the property as an earnest of sincerity at the time of a reconciliation.
>
> The Prothonotary found that the story told by the wife and her witness was correct. The application was dismissed in respect of all articles except a leather motor coat which had been returned to the husband.

Ellen and her witness. Who else but Rose would insert herself into the Macbeths' public bickering? Returning to my 1980 notes I can now decipher a sentence that made no sense at the time: 'Jane bought that car from us.' Jane,

William's rescuing sister, pitted against Rose, Ellen's self-appointed guardian. Jane in damage-control mode, buying the car with her own money; Ellen (and Rose) taking the money because they had to — and William, for all we know, being given the car back (with conditions).

Not quite seven years after this court case Ellen suddenly died. She was thirty-three. The cause of death, coronary occlusion, was in Rose's opinion just a medical diagnosis. My father was ten years old and apparently orphaned (William hadn't been heard from in years). What happened in the aftermath of Ellen's death marked my father and his brother in permanently disorienting ways. The boys had to be 'placed' somewhere. They couldn't stay with their new stepfather, and they couldn't be grafted into the cousin network of Ellen's sisters (both of these 'couldn'ts' have never really been satisfactorily explained); there was categorical agreement that no approaches were to be made to the Macbeths, although they were a prosperous family, and we know that Jane at least was benevolently inclined to William, and William's doting mother was still alive (she outlived William).

The placement, when it happened, seems to have embarrassed Ellen's family into silence. The boys were 'given' to Mrs C, a woman in her sixties who needed workers on her few acres of farm. It sounds Dickensian, and it undoubtedly *was* from the brothers' point of view as they laboured long hours for their board in a void of affection or physical reassurance; but it isn't fair to demonise this woman, or to judge her attempts at instilling a work ethic into the young boys as anything other than adapting the means to the end. After four years of this sunrise to sunset labouring, my father ran away to the country looking for his family. It was a sobering experience. He saw the poor holdings and make-do existence of rural insolvency, and discovered a chaos of half-finished projects abandoned to apathy that threw a strange, forgiving light on Mrs C's bountiful three acres.

Propping his fourteen-year-old self against the railing of his mother's hometown pub one day, my father was surprised by a man telling him he was the spitting image of William — his unmentionable, murderous father. The likeness, he was told, was uncanny. The source of this comment, the local solicitor, smiled affectionately at the memory of William Macbeth setting their little community alight back in twenty-one or twenty-two with his stage show, marrying one of the Clark girls and taking her to the bright lights of the city. Where, the solicitor was interested to know, did Macbeth live now?

Unable to assimilate this image of William into anything in his experience, my father pretended mistaken identity. Part of him wanted to listen to more stories — it was obvious there were more to come — but the greater part was too infected with antipathy.

He left on the next train and returned to the less questioning world of Mrs C.

>―◦―◦―◦―<

Irrational fears are a failure to integrate an experience (usually a negative one) in the normal way.

I don't ever want to be poisoned and don't expect to be, yet I have what might be termed an 'iophobic respect' for cyanide. This respect is not grounded in personal experience, yet the circuitry was laid down at some point in my development, and I can recognise the symptoms.

Fear is a strange beast: it rears up at both the known and the unknown. If you listed the top ten deadly toxins, and told me that an assassin's injection of potassium chloride would give me a quick, fatal heart seizure; or that in a depressurised plane I'd become hypoxic after a brief euphoria and drowsiness, then pass out and die, I might thank you for the information, but the point would be academic, the fear beast would sleep through it. If on the other hand you dropped a Zyklon-B pellet into my shower cubicle, the terror might kill me before the cyanide.

Every pharmacy student in my time knew the story of the female chemist who drank water from a glass measure and dropped dead because, clinging to the sides of the glass, invisible, was a residue of the last liquid it held, strong cyanide solution. The story annoyed me at the time, and still does today, because of its gender politics and because, if nothing else, chemists are trained to use their noses as well as their common sense. But apocryphal stories, like fables, catch the attention. This story has stayed with me, and may be the one that seared its singular imprint.

Cyanide has its own cache of stories, all characterised by the element of speed and surprise. An early scientist said of cyanide, 'after a fatal dose no voluntary act of significance — save, it may be, a cry for assistance — is performed.' Inhaling the vapour brings on unconsciousness in seconds, death in minutes. The drama of cyanide is part of its attraction to the suicide who doesn't care a fig what the body looks like after death. (There is some speculation that Cleopatra dismissed the cyanide solution because of its tendency to turn the mouth, and sometimes the whole face, bright blue.)

In the goldfields of Western Australia, where cyanide

was once as plentiful as salt, a miner accused of embezzlement asked the court if he might smoke a last cigarette before judgement was pronounced. Rising to his feet, the accused extracted a hand-rolled cigarette from his pocket, faced the jury, lit up, inhaled a lungful of cyanide-laced tobacco, and died on the spot. Counsel for the defence scattered to avoid fumes rising from the fallen cigarette.

Cyanide is absorbed from all tissues except the intact skin. A spill on the hand might only require careful rinsing. A spill on a cut finger might mean quick death. When an excessive oral dose is taken there is salivation, and a series of very unpleasant choking sensations, culminating in a mouth foaming with almond breath, and collapse. The interval between the dose and onset of symptoms can be as short as ten seconds. If the dose is large enough the intermediate stages merge into one another and can't be distinguished. The person is seen to stagger a few paces and fall, like the embezzling miner.

In the 1860s a doctor researching cyanide asked himself: What can a person in full possession of his faculties do in the ten seconds before unconsciousness? The doctor, stopwatch in hand, drank some harmless liquid, corked the bottle, threw it out the window, got into bed and arranged the bedclothes tidily. Exactly ten seconds, an experiment that had medico-legal as well as empirical significance. (Adolf Hitler was able to point a gun at his head and fire an accurate shot.)

As always, there are qualifications to the dose–response relationship. The lowest concentrations of prussic-acid-forming substances are found in certain fruit kernels — like peaches, cherries, apricots — and in almonds and cassava roots. The high-grade, high-potency form of cyanide, the chemical salts, come in packets and tins from the manufacturer. The more pure the cyanide, the faster the effect.

A woman who drank a bottle of essence of almonds (less pure than refined cyanide) had time to go to the well, draw water, drink, climb up two flights of stairs, call her child, come back down the stairs and fall dead on her bed, thirty minutes from start to finish.

A maid who quarrelled with her lover ate a large quantity of bitter almonds (less pure again), passed out after ten minutes and died after ninety.

At post-mortem, these women appeared to have died of suffocation, and in the early days of determining cause of death the pathologist was advised to open the head first, where a strong smell of almonds would differentiate between cyanide and choking.

Given these variations in response times, my irrational self would rather be handed a cyanide cigarette than a bag of bitter almonds.

⊱─◆──○──◆─⊰

Capital punishment using cyanide goes back to antiquity, and persists today in some states of America where death by lethal injection, or gas, is the option favoured over hanging.

Thousands of years ago Egyptian law-enforcers introduced what they called 'the penalty of the peach' for such arbitrary crimes as saying secret words aloud. The Romans copied the method. Peach penalties, and suicides, became an alternative to the sword. In the reign of Tiberius, a high-ranking officer accused of treason staged a dramatic suicide at the feet of the senators by draining a cup of peach-sourced cyanide.

As an instrument of execution, cyanide is considered to have many attractive and useful advantages over competitors. The results of trials of a pesticide, Zyklon-B (cyanide in pellet form), in Nazi concentration camps in World War II confirmed that six kilos of Zyklon-B despatched 1492 prisoners in less than five minutes, the people closest to the inlets succumbing in seconds, those protected by distance from the source and density of people dying last.

Two years after World War II ended, four men executed in the United States by inhalation of lethal doses of cyanide were studied by scientists who connected them up to continuous electrocardiograph tracings. After the men were strapped in the execution chairs the scientists attempted to obtain 'control' traces, to compare with the traces after the gas was released. The men exhibited 'normal' baseline heart rates of between 102 and 166 beats per minute, which were abnormally high and indicated, I imagine, that they were very frightened.

After the gas was released the electrocardiograph traces jumped all over the place. At the start the heart rate slowed down and the rhythm changed. There was a speeding up in the third and fourth minutes, then a slowing down in the fifth. The speed increased again in the sixth and seventh minutes, then slowed, until each man's heart failed in a different way yet continued to beat for several minutes after the last breath.

Scientists have been watching experimental doses for years, stopping short of lethal jabs. When I was a student you could earn extra money by drug-testing, and in class you could participate in drug-taking experiments, like swallowing a diuretic pill and measuring your urine output over a given period. Universities pay students to sip poison and let themselves be watched. In a 1947 study, healthy volunteers let themselves be injected with small measured doses of cyanide. The results are not unlike the patterns of the four executed men — erratic heartbeats and breathing — except for the endpoint when their hearts returned to normal because their young healthy bodies had coped. In fact repeated daily injections of these subjects with cyanide 'to the point of unconsciousness and convulsions' left the nervous system 'in good order'.

Experimentation is an attempt to take control of the variables, to make poison and flesh obey science's rules, to banish the accidental and the fortuitous. Calculating how many grams of cyanide kill how many prisoners in how many minutes in a sealed chamber is a cruel debasement of the spirit of scientific investigation, but is not unfaithful to the practice. The practice requires its bell jars and killing bottles, its men with their measuring instruments and their unflinching gaze.

Chapter Twenty-one

Murder on the table

THE SUGGESTION THAT William Macbeth *executed* his sons insinuated its dark thread into the fabric of my grandfather's history in a desperate bit of last minute stitching. Under my pressure, Rose's originally pliant accusation against her brother-in-law recast itself into a garment with fortified seams.

Any dictionary will give you at least two meanings for 'execute': its everyday, commonly understood sense of carrying out an action to a plan; and its juridical sense of putting to death according to law. Execution is the boldest of all acts of will against another human being. The victor wields his power over the vanquished; punishment obediently follows crime. But my imagination would have to stretch a long way indeed to accept the picture of my grandfather entering into the solemnity of a ritual-like execution to eliminate his sons. Macbeth manifested many idiosyncrasies, many deficits of character, but I doubt he had the exact degree of coolness in his blood to carry such an act to completion.

Nevertheless, the proposition is on the table and, in view of the serious nature of its implications, I want to look at two execution stories (one old, one new, and both well known) and line them up with William's story in a crude attempt at cross-matching.

⊱—⊰⊙⊱—⊰

My first subject is Socrates, executed under the laws of Athens in 399 BC by the (to modern eyes) unusual method of drinking a cup of poison. I first came to the death of Socrates through Plato, as most readers do, and carefully followed the dialogue and description (this is a wordy death scene) so that I too could arrive at an understanding of the principles behind the philosopher's unswerving determination to die well. Later I stood in front of Jacques-Louis David's 1787 painting, *The Death of Socrates*, in the Metropolitan Museum of Art in New York, and experienced the same perverse pull that my reading produced — that is, a disinclination to take my eyes off the cup of hemlock. This non-Socratic reaction reflects, I'm sure, a life spent watching the work of drugs in bodies, and puts me in the category of the pedant who can't see past her own learning. But as this journey concerns itself not with philosophy but with the physicality of poison, I begin my examination with my gaze directed at the cup.

David painted the scene in a neoclassical style that typified the sentiment of the age: civic virtue, austerity, noble and heroic themes expressed economically. Seventy-year-old Socrates, robust, muscular, seated on a bed with one leg outstretched, the other still touching a foot-stool, holds forth, gesturing with his left hand, while his right hand hovers above the execution cup, centre-screen. Hemlock, the ancients believed, granted a painless death. The dying started from the feet and worked upwards, gradually extinguishing the lights as it climbed the stairs, a ponderously slow process in the eyes of many, but not a process, one suspects, that Socrates had any quarrel with.

In various poses around Socrates are twelve people (count them and consider the tolling of that number, twelve in a jury, twelve apostles, twelve months in a year, and so on). One is his grieving wife, who is shown leaving the scene, withdrawing her feminine contribution to mourning — tears — from the fine, intellectual moment. To the far left, in the foreground, slumped on a chair at the end of the bed, is the old man Plato (the chronology is acknowledged to be wrong: Plato was in his twenties when Socrates died). To the far right is a figure who is, reputedly, a physician who has mixed the hemlock according to his skill, and is passing by on his rounds, impervious to the patient/prisoner on the bed. The death cup, a rather commonplace vessel, is offered to the prisoner (who is too absorbed in conversation to notice it) by an unhappy young man going through the motions of doing his job.

The Death of Socrates is, to me, a disappointingly static canvas from an artist who painted another death scene, *Marat Assassiné,* with such expressive pictorial eloquence. Sir Joshua Reynolds is supposed to have called the *Socrates* 'perfect in all parts'; others disagree (with both the sentiment and the attribution). I find it lacking in involvement, as if the artist, who we've been told was a man of action and political energy, was entirely out of sympathy with the philosopher's attitude of calm acceptance.

This painting, I've since learned, is used in the medical humanities (in particular, palliative medicine) as a stimulus to discussion. What do you see, what do you infer? workshop participants are asked. For instance, is this a representation of a 'good death', the goal of other death-seekers like Cleopatra or Emma Bovary? Who is the man with his hand gently touching Socrates' knee? Is he the calm witness, the quiet escort waiting for his cue? The group answers are interesting because they underline the great differences in point of view among people who 'read' a scene. Some groups are confused (as I was) about which figure represents Socrates. Is he the slumped (dead? meditating?) figure at the foot of the bed, or the robust gesticulating central figure, getting in those last words?

There is great historical and philosophical value in having a first-hand witness account to the final moments of an execution, but we have to entertain some healthy suspicion about the 'easy' time Socrates had in dying. My chemist's nose wants to sniff the potion in the cup, to test the theory that the hemlock was spiced with henbane and opium, a combination that might offset some of hemlock's more unpleasant side-effects.

Plato's account of Socrates lying calmly as the coldness spread gradually upwards suggests that the process was at least bearable. The great man covered his face, maybe to hide a grimace or two, as he concentrated on reaching a tranquil state of mind for his smooth passage to the next world.

Crito is probably speaking for all concerned care-givers when he tells Socrates, 'the man who is to give you the poison has been asking me for a long time to tell you to talk as little as possible; he says that talking makes you heated, and that you ought not to do anything to affect the action of the poison. Otherwise it is sometimes necessary to take a second dose, or even a third.'

Socrates answers: 'That is his affair. Let him make his own preparations for administering it twice or three times if necessary.'

The affair that occupies Socrates is dying *by example*. The physical symphony that brings him to this point is unimportant, even uninteresting, and involves matters better delegated to experts.

Sliding an imagined scene of William murdering his sons into place alongside *The Death of Socrates* might seem a little far-fetched, but what can be learned from 'reading' the David painting applies just as well to an obscure execution. My eyes can compose the three figures in a painterly way: Macbeth, the executioner, oblivious to the feelings of his victim, measures the fatal dose and holds the cup for the child to drink. The boy, on his bed, is looking not at his father but at his brother who is playing nearby. Smiling, he feels the cup meet his hand, and the moment freezes, like a photograph or a painting.

The analogy works at the level of logic at least: there is an inference of similarity from the similarity of two or more things in each image: the trusting unknowingness of an innocent child against the trusting knowingness of the adult; and central to each scene, poison in a cup. In both cases execution is possible.

The other image on my light box is the beetle-browed, self-conscious murderer, Graham Young; a callous executioner of family and friends.[45] There is no temptation to elevate any of Young's deeds to a philosophical plane. He was, his biographer tells us, a disturbed child with an abnormal interest in poison and Nazism, who at thirteen began to manifest the obsessive, oddly inhuman behaviour of the sociopath. In 1962, aged fourteen, Young poisoned his stepmother (who died), his schoolfriends, father and sister (who lived), and was sentenced to Broadmoor psychiatric hospital for the criminally insane. Released through the intervention of a psychiatrist when he was twenty-three and placed in a good job through an ex-prisoner support scheme that guaranteed anonymity, Young resumed his poisoning career, killed two workmates, and was put back into gaol where he died in 1990.

Recently I watched a film about Young, *The Young Poisoner's Handbook*, a black, even bleak, comedy that re-orients the viewer (it feels as if the chairs have been rearranged, so that Young himself mediates what is seen and known). In life, Young spoke in a curious, formal voice, and through the medium of the film we feel his cold detachment from normal human relationships. What kept me watching was a sort of fascinated horror for the arch poisoner at his work, with a dose in one hand and a notebook in the other. Young kept meticulous notes on his favourite poisons (antimony, thallium, belladonna), his favourite poisoners (Crippen and Palmer), and of course, his own evolving skills at the business of being a good poisoner. Young played a cat-and-mouse game with his victims, bringing them to the point of death, then backing off. Portions of his notes read, 'D's loss of hair is almost total . . . it looks like I might be detected.' And 'F must have phenomenal strength. He is being obstinately difficult.'

Statements made by Young at his first arrest in 1962, and the second in 1971, reveal him boasting to police and psychiatrists about his knowledge of poisons, correcting their mistakes and, in one instance, requesting they give him back his antimony because, he said, 'I miss the power it gives me.' His early notebooks show a mind fascinated by the mechanics of death; experiments on mice, cats, schoolfriends and finally his immediate family. When he was poisoning his workmates at the opticals factory, Young became impatient with investigators and began airing his knowledge on heavy metal poisoning, stressing the relationship between loss of hair and thallium overdose. He literally gave himself away to save the tedium of waiting for amateurs to catch up with his cleverness.

The screen image of Young somehow in thrall to poison, seduced by its lure, smacks shamelessly of Dr Jekyll struggling to resist drinking the potion that will turn him into Mr Hyde. There is a scene in which Young is shown around the laboratory of his new (post-prison) workplace, and a cupboard is opened to reveal rows of bottles of thallium. Gazing at the hoard, Young's face lights up — one glance and he's 'hooked' all over again, the old demons wake up, and we, who are inside his head, know that the inevitable is about to be made manifest.

How many matching characteristics tie Young's execution style to William Macbeth's? Macbeth had a knowledge of, but not an *obsession* with poisons. Macbeth was a dabbler. Nevertheless, he dabbled with strychnine, and possibly other restricted substances, in the context of experimentation, and this is a partial match with Young. Turning to the victims, in both cases each man poisoned at least one member of his own family. In the realm of generalities, both men died prematurely, in their forties, a match that can only be laid at the door of coincidence by any measuring stick.

The greatest area of disagreement comes after the

events. Young was completely devoid of the rescuing faculty. As his stepmother died by inches he exhibited the deafness and blindness of the vivisectionist to animal pain. Macbeth, on the other hand, called a doctor for his stricken son.

In neither of my test cases is the rescuing faculty switched on: Socrates categorically rejected rescue, Young in his pseudo-autistic way failed to connect with it.

I want to turn away from these cold-blooded men to look at the rescuers.

Chapter Twenty-two

Rescue

IF EMMA BOVARY (no longer conscious and deep in the throes of arsenic poisoning) had been rushed to the emergency room of a modern teaching hospital she'd have collided with all the dazzling medical resources of the late twentieth century. Propelled on a trolley, bursting through doors, Emma's body would suddenly arrive at a standstill under bright lights, like a racing car swinging into the pits, to be worked on by a team ready to probe her parts and apply their science to each.

Oblivious, Emma and her mortal illness would become the nexus of a closed loop that passes in and out of the body on the table.

Arsenic is a rare visitor to the ER these days. The younger staff won't have seen it before; an old hand might remember it from the sixties, but even the most educated guess is subservient to the authority of the tox-screen. While Emma's blood is assayed, the team apply the well-rehearsed poisoning protocol: assess cardiac and respiratory status; get rid of any unabsorbed poison by purging and washing the stomach; apply the antidote aggressively (if no test results are back, use a slurry of activated charcoal); support the collateral damage of pain and shock with fluids and morphine. If all goes well, she might pull through. It's a matter of timing and persistence.

⤙⦾⤚

When Charles Bovary finally comprehends that his wife has swallowed arsenic he stumbles about the room, knocking over furniture, calling out, 'Poisoned! Poisoned!' Completely incapable of acting in any medical sense, he writes desperate appeals for help to his colleagues then turns the pages of his medical dictionary in a futile attempt (his eyes won't focus) to read up on the poison's actions. The maid fetches Homais, the chemist.

> 'Keep calm!' said the chemist. 'It's only a matter of administering some powerful antidote.'
>
> [*When told the poison is arsenic, Homais says,*] 'Right! We must analyse.'
>
> For he knew that in all cases of poisoning it was necessary to analyse. Charles, not understanding, answered, 'Yes, yes, go on! Save her!'

Antidotes are strange and wonderful entities. Some have always been with us; others have been purpose-invented to fill the gaps. In Emma's time, the closest thing to an arsenic antidote was an iron mixture which none of the attending medical men — including her husband — thought of, or managed to obtain. The textbook treatment for arsenic poisoning in the 1850s is an eerily similar, if raw, version of what happens today:

what is different, and finer, is the clever piece of chemistry that produced a specific antidote.

The word 'antidote' derives from the Greek *antidotas*, meaning 'given against'. In medical usage, an antidote is a substance that stops poison doing its work. The word has a positive, rescuing connotation and its synonyms — counter-poison, mithridate, antivenin, antitoxin, remedy, cure, neutraliser — line up beside the original like patriot missiles waiting to be launched.

Like other 'anti' words — antibody, anticoagulant, anti-tank, antipathy — one thinks of battles, campaigns, mind games; effort that has gone into studying the enemy, learning its ways, exploiting its weakness.

Quite a lot of the business of feeling safe in our everyday world relies on our belief in remedies. The idea that antidotes are stored in surgeries, hospitals and bathroom cupboards is a comforting 'given' in modern Western society. Psychologically and historically we've developed a dependency on the idea that the harm done by toxins can be undone by a counter-agent whose accuracy and reliability has been guaranteed by science.

In the late nineteenth century when poisoning was commonplace and 'all medical men in practice were liable to be summoned hastily to a case', up-to-date doctors carried an 'antidote bag'. The bag contained a stomach pump — usually 6 to 8 feet of India rubber tubing — a hypodermic syringe, a bleeding lancet, a tracheotomy knife, a small 'interrupted current' battery (to supply galvanic shocks to the walls of the chest); at least four different emetics to induce vomiting, and a careful selection of antidotes, such as they were.

Dr Blyth, author of this list, is very specific, and confident, in his choices. For additional ammunition, he carried a flask of brandy, ether, smelling salts and some 'really good coffee extract', the latter to administer as an enema or stomach wash.

>-+◆>-O-<◆-+-<

At Emma Bovary's bedside the wheels of nineteenth-century science hardly seem to turn at all.

Canivet, a Doctor of Medicine, arrives. In the unique French medical hierarchy of the time, the doctor was a cut above Charles Bovary (a mere Officer of Health). Canivet prescribes an emetic 'to clear the stomach completely'. Charles and the chemist dither about the bedside, waiting for Canivet's miracle to work. Instead, Emma gets worse. In an extraordinary lapse of compassion or attention Canivet prepares, *but stops short* of giving Emma the antidote. The moment passes because of the flurry caused by the arrival of yet another medical master, the fabled surgeon Dr Larivière — who takes one look at Emma and knows she is doomed.

Swallowed arsenic scours the alimentary tract on its way to the stomach. The capillary beds along the route swell and burst, the cell plasma collects into blisters under the membranes, the blisters eventually rupture. But compared to the damage that comes later, the scouring is like a superficial graze. The irritant drug passes through the gut membranes into the bloodstream, and like a pollutant released into the headwaters is carried away on currents that wash up their debris against incidental barriers.

Not all the coarse powder swallowed by Emma dissolves. She vomits a white sediment into the bowl Charles places at her lips. The portion that does dissolve will react with her body fluids, change its chemistry, and pass into the liver, kidney, lungs, and, even some weeks later, *after death*, into muscles, nerves, hair, nails, bones and teeth. If Emma had been pregnant, it would have passed into the placenta, and her breast milk.

Arsenic is not fussy about the body being alive. Cases of 'cadaveric imbibition' have been cited. The famous toxicologist Joseph Orfila, writing in 1814, describes introducing arsenic into the stomach and rectum of a dead man (in the interests of science). He made dissections after eight, ten and twenty days. The arsenic had moved from the stomach to the liver, the diaphragm and lungs.

>-+◆>-○-<◆+-<

Emma was lucky to pass through her ordeal relatively quickly. Others have taken longer at the business of dying. Ten years before Emma, on a Wednesday in August 1847, the Duc de Praslin took a suicidal dose of arsenic. By Friday, after the usual vomiting, diarrhoea, and thready pulse, 'great coldness of the limbs' set in, and (understandably) depression. On Saturday he felt better though he'd ceased passing urine and felt feverish. However, things changed dramatically on Sunday. He became extremely thirsty and had a burning pain from 'mouth to anus'. Between Sunday and the small hours of Tuesday morning when he finally died (six days after his fatal dose) he endured an erratic pulse and respiration, and thrashed about so much his carers got no sleep.

Discrepancies in outcomes, like Emma's and the Duke's, or Frau Erpenstein's, or anyone from the long rollcall of arsenic deaths, are poised on pharmacological facts about what arsenic does to tissues, facts that have had successive layers of understanding pasted over the original and where old knowledge is still visible under the new.

Sir William Henry Willcox threw out the theory that arsenic killed by inflammation (and therefore internal bleeding) once and for all in 1922.[46] 'Arsenic does not kill by virtue of its *irritant attack* on the gastro-intestinal tract (although it is certainly very irritant); rather, arsenic is a most powerful tissue poison, and it is to the toxic effect on the important organs of the body, especially the heart, kidneys and liver, that a fatal result is due.' His illustrative case history concerns an unfortunate man who mixed white arsenic powder into a glass of hot milk, thinking it was his usual evening antacid. He followed this with a supper of porridge then went to bed. Four hours later, at midnight, he had burning epigastric pain and then an episode of diarrhoea. At 4 a.m. his heart began to fail, his pulse was thready, and he had agonising cramps in his legs. At 10 a.m. he could no longer pass urine, and seemed beyond all help. But, reports Dr Willcox, *with treatment* the acute symptoms subsided and he was pronounced recovered from his ordeal.

But was this the end of story? No.

Dr Willcox pauses for effect, before delivering his *coup de grâce*. One hundred days after his arsenic and milk drink the man in question died from severe and fatal nerve damage, directly attributable to the arsenic; the time interval being how long it took the poison to infiltrate and decimate his nervous tissue. 'This case is a good example of the action of arsenic, which is that of a protoplasmic poison, the irritant effects not being *of themselves* fatal.'

'When your forces are dull,' writes Sun Tzu in *The Art of War* 2000 years ago, 'your edge is blunted, your strength is exhausted, and your supplies are gone, then others will take advantage of your debility and rise up. Then even if you have wise advisers you cannot make things turn out well in the end.'

The maxim for treating arsenic poisoning, 'always give an antidote, be the case ever so hopeless', makes sense when you know that the enemy continues to march even if its opponent is dead. Emma Bovary, still clinging to a bit of life when her putative saviours gather, might have rallied with some expert help. But it was never Flaubert's intention to spare Emma. In a letter to his friend Bouilhet he wrote: 'When I was describing the poisoning of Emma Bovary I had such a taste of arsenic in my mouth and was poisoned so effectively myself, that I had two attacks of indigestion, one after the other — two very real attacks, for I vomited my entire dinner.'

Knowing what to do, and when and how to do it, is as important as finding a counter-poison. Antidoting did not arrive in the twenty-first century in handy vials or bottles ready to use, but grew, like any branch of medicine, by extrapolation from folk cures, happy accidents and inspired chemistry. Some poisons have no antidote at all (the toxin produced by Australia's blue-ringed octopus, for instance), *two per cent* have a specific antagonist, and the rest, the other nearly 3000 toxic substances that threaten life, have to be outsmarted by strategies, sometimes as drastic as exchanging poisoned blood for fresh blood by transfusion; sometimes as cautious as waiting and watching, with no intervention except keeping the body warm and topped up with fluids.

Dimercaprol was purpose-invented in 1945 as an antidote to arsenic war-gas. In a brilliant bit of chemistry a molecule was developed which could grasp the toxic metal and hold it in a ring or bracelet. The new molecule (dimercaprol linked to arsenic) would then be sent packing, carried away along one of the body's normal excretion routes, the urine. The drug, given by deep intramuscular injection, at different sites, and in divided doses, routed the enemy, but not without some cost to the army: some unfortunate patients have died from the over-energetic application of the cure.

Western medicine is very much taken up with the lock-and-key picture, and this single-minded approach has achieved startling direct hits. Naloxone, given intravenously, can reverse an opiate overdose in two minutes; flumazenil reverses benzodiazepine intoxication; atropine rapidly blocks the paralysing effects of organophosphates; acetylcysteine is the antidote of choice for paracetamol overdose; and Digoxin Immune Fab (or Digibind), the last-resort but very effective antidote to digoxin overdose, has been shown to block at least two other poisons that shock the heart out of its safe rhythms: toad venom and yew berries. (Of six males who ate 'love stone', a Chinese aphrodisiac made from toad venom, four died of heart failure, but two were saved after infusion of ten to twenty vials of antidote. A five-year-old girl survived the nearly fatal outcome of eating berries from a yew tree, after cardiac resuscitation and pacing, and infusion of Digibind.)

Instinctively, in the face of illness, accident, poisoning, we go into rescue mode. Like Isis, we try to stitch the dismembered Osiris back together again. The scramble to swallow something in order to 'normalise' an abnormal sensation prevails; acute discomfort is like a vacuum, abhorred and quickly filled.

Osiris was murdered by his brother Seth, his parts scattered far and wide, lost apparently to all except his wife Isis who became the tireless gatherer of the fragments of her husband's body. When she had all the pieces she wrapped Osiris in her gently beating wings (or lay on his body depending on who you read) and breathed life back into him. By whatever magic or knowledge or divine trickery, Osiris was rescued, and rescue has become one of our most potent social metaphors.

>-+-+>-+O-+<+-+-<

Here is my own rescue fantasy for three-year-old Patrick Macbeth:

Visiting Australia in 1927, *en route* to a lecture tour in London, is a world authority on arrow poisons, Dr John D. Gimlette.

He's travelled from his hotel in Sydney to the mountains for a change of air, and to spend a few days with his old friend Dr Allan. They were at school together in England, and although Allan's modest general practice might suggest a congealed interest in discoveries made in distant lands, Gimlette has found Allan to be an appreciative and thoughtful correspondent. Gimlette's book *Malay Poisons and Charm Cures*, in its second enlarged edition, has been reviewed in a slightly tongue-in-cheek manner by the *British Medical Journal* and it was Allan who penned the well-argued riposte in the Letters Column.

It's August, and bitterly cold. Gimlette and Allan have been up half the night reminiscing, and now, after Allan's morning surgery, they're packing their kits for a half-day hike in the Megalong Valley. Allan takes few afternoons off, so the day has the air of a special occasion.

But before they can escape, the telephone rings and they hear Allan's wife telling the caller to stay calm and repeat the message. Even while tying his shoe and chatting with Gimlette, Allan is listening for trouble.

The door opens and Allan's wife holds out the phone with a gesture of regret. She shrugs in Gimlette's direction. Allan takes the call. It's a frantic mother. Her son is having a fit.

Allan makes a note, rings off and frowns to himself. He tells his wife to call the ambulance and to ask their teenage son, who will relish the task, to warm up the car and back it onto the street, quickly.

Allan doesn't know the family concerned personally but he's heard about the husband on the local grapevine. Apparently Macbeth sells patent medicines. He's a colourful character, wears a morning suit, works from home. Allan has no time for these shysters and their snake oil. He apologises to his guest.

'Wait,' calls Gimlette and hurries to catch Allan, who's already at the car door.

On the short trip to Wilson Street, Gimlette remarks that he's spent twenty years with savages. A city shyster could only be tame by comparison. If Allan doesn't mind, he'd like to look over his shoulder.

At the Macbeth house a small crowd has gathered inside the gate. An onlooker who knows Dr Allan begins to summarise events. In her opinion, the boy's been poisoned, and quite frankly, she's *not surprised*. Allan, followed by Gimlette, enters the house, closing the door firmly behind them.

It's a typical mountain house, solid, with a lot of small rooms that can be closed off. They follow the central corridor and turn into the sitting room where a woman's hysterical crying locates the scene of the drama. Macbeth, who is leaning over the boy, straightens up when they enter and puts out his hand. Allan shakes it, introduces Gimlette and gets straight to work.

'He was sitting beside the fire,' says the wife, 'singing.'

Clearly the boy is very ill. Allan decides to administer an emetic. Gimlette begs a word in private.

'I can tell you unequivocally,' says Gimlette, 'the boy's taken strychnine. I see it almost daily in the jungle, the natives use a crude form on their arrows.' He reminds Allan that a paralytic poison should never be treated with emetics once the convulsions have started. The boy has fitted once and as they speak is going into spasm again.

'There's not a minute to lose,' says Gimlette.

Allan gladly defers to his colleague. He's out of his depth and has been listening for the ambulance.

Gimlette orders everyone from the room, the light to be turned off. 'We must have absolute and total quiet, not even a footfall.' To the mother he says, 'I need all the pillows you can find.' To Allan he says, 'Do you have ether, or chloral?'

Macbeth speaks. 'I have pure chloroform.'

'Then bring it immediately.'

The men exchange a glance. Macbeth brings the bottle. He stands to one side of Allan, watching Gimlette work. 'The trick,' says Gimlette, 'is to keep the boy as still as a mouse. This can take hours.'

Allan goes outside to explain the situation to Mrs Macbeth, and to escort her and her small sons to the neighbour's house where she must wait quietly, and with *hope.*

>―←→―•―←◦―→―◄

The ancients gave hemlock against strychnine's clenchings. Nineteenth-century medicine gave anaesthetics and choreography:

> Place the patient at once under chloroform or ether, and keep up gentle narcosis for several hours, if necessary. Darken the room, stifle all noise; if in a town, and opportunity permit, have straw or peat placed at once before the house to deaden the noise. Chloral hydrate may be used in the place of chloroform, even the juice from a recently smoked pipe diffused in a little water and injected subcutaneously might be tried.

Today, the drug that unclenches strychnine's hold is Valium, and a machine does the work of the lungs. Otherwise the steps are the same: 'All unnecessary external stimuli should be avoided and if possible the patient should be kept in a quiet darkened room.'

When dogs take strychnine baits (and can be got to a veterinarian in time) the surgeon injects sodium pentobarb intravenously to the point of anaesthetising, and re-injects at intervals as the anaesthetic wears off. The dog may be kept in a state of suspended animation for up to seven days while the slow excretion of strychnine takes place.

The real Dr John D. Gimlette trenchantly records the way jungle Malays treat accidentally self-inflicted arrow poisoning (the poison being a complicated mix of ingredients, often including vegetable strychnine).[47] 'One day,' he writes, 'a poisoned blowpipe dart fell out of a quiver and stuck in the upper part of one of the men's feet. It was at once pulled out, and a Semang squeezed the wound to get out as much blood as possible, then tied a tight ligature round his leg, and put lime juice on the wound.' Lime juice, human urine, a mouthful of dry earth, or the fruit of an onion-smelling jungle tree are the only antidotes the jungle Malays give any thought to.

The simplest antidotes are often found somewhere in a natural state — a leaf perhaps, or a mineral in the ground. Some turn of events leads to a discovery of its special properties which are further accentuated by refinement (usually in the West); or the plant or mineral is left in its raw state and 'worked up' on the spot when needed (usually not in the West).

The antidotes we choose, or have chosen for us, are fashioned by all sorts of practitioners, each one admissible on its own evidence.

Traditional allopathic medicine takes cases of poisoning straight to the emergency room, or the closest doctor or chemist. This mechanistic approach locates the drama in the disrupted physiology of the body, a location that can be assessed and decommissioned by the application of accurate, powerful counteractive medical attention.

Other groups throughout history have dosed themselves *in anticipation* of toxic attack. By training and supplementing natural immunity the believers in prevention-over-cure hope to shield themselves with discipline.

The so-called 'mithridates', strange concoctions of

multiple ingredients said to counteract every known poison, including snakebite, have been invented and re-invented since the days of Mithridates, a king who lived in constant fear of being poisoned. His physician put together a general antidote of forty or more ingredients which the king took in alternating doses with different poisons until he developed immunity.

In the presence of actual (as opposed to projected) danger, as for example, during the terrible plague years, careful citizens held warm, fresh bread in their mouths to absorb foulness (my mother-in-law does the same thing today when cutting onions). Construing their enemy to be in the air these same citizens used livestock as primi-tive air-exchange units: a cow in the house to exhale sweet air; or a billy-goat to overpower the bad air with even worse breath. Air-borne toxins, the so-called 'mephites', were thought to rise up from swamps, mines, caves, pits in the ground, and escape in the breath of a plague victim.

Dreckapotheke, from a Dutch word meaning dirt or filth pharmacy, is an approach to illness (acute and chronic) still used by societal groups who embrace the supernatural. The literature records many of these bizarre treatments, which often involve excrement. To this record I add some examples brought back from India by a friend who lived for ten years in a remote temple high on a mountain in the desert. In her former Australian life she was a nurse, trained in allopathic medicine, antibiotics and autoclave sterilisation, but now, after so long in self-imposed exile, immersed in the prevailing ethos of temple life, she's reluctant to revert to her former trained beliefs.

If she brought someone to the temple with an infection or poisoned sore that was the result of a flesh wound — no matter how small — the priest blamed demons, not bacteria, and a lecture would ensue. Blame automatically reverted to the individual who had done something 'wrong'. In the supernatural cosmology of the temple unseen forces enter the body when the protective sheath of the skin is breached — the hole acts as a conduit for a demon to get *in,* and a portal from which vital energies can slowly leak away. The demon has to be exorcised with a treatment ugly enough to drive it off; this would often be dried rat faeces and ground turmeric mixed into ghee and poulticed onto the sore spot. (Puncture attacks by snakes, centipedes and scorpions are manifestations of the striking, reckless nature of demons and can only be warded off by mantras and peacock feathers.) A good dousing with cow's urine will disinfect the site when the crisis has passed; and if the demon reappears, in the form of a recurrence of symptoms, it can be drawn out of the mouth by dabbing the tongue with a mixture of cow's milk, urine and faeces combined with yoghurt. If there is contagion in the air, an incense stick made from equal parts of dried donkey and dog faeces with pieces of human hair (another variation of cow and billy-goat breath) can be made to slowly smoulder, and the fumes inhaled. The work of defeating poisons and driving off demons is often an inherited gift, guarded by a keeper.

>-+-+>-O-<+-+-<

The least acknowledged, yet best standby approach, to antidoting is what I call the 'tea and toast' school, an empirical domestic domain that draws on the wisdom of the kitchen. It assumes the absence of expert help, because that is always the first port of call, and demands a good measure of common sense. In *Shadows on the Grass*, Isak Dinesen — whose African farm was miles from anywhere — hurried to the aid of a Masai boy poisoned by Lysol, carrying bicarbonate and oil which she used 'against accidents with corrosives'. At another time, when Dinesen accidentally took an overdose of arsenic drops (her syphilis medicine) she reached for an Alexandre Dumas novel for its description of a king saved by milk and egg white.

In ancient times and other worlds, away from the classifiers and their laboratories, healers who were consulted when the wrong toadstool or a doubtful dose of hemlock was eaten, went to their gardens to pick herbs which, when mixed with wine or vinegar, were the match of any poison.

Herbal infusions, according to this ancient school, can cure the bite of the shrewmouse and the sea-dragon; and perhaps they can.

━◆◆◇◆◆━

Here is a rescue fantasy to save a doomed king:

In the court of old King Roderick, the greatest Dane of all, according to some, any matters concerning the diet of the king (whose digestion was faulty and was said to account for his temper) were referred to Bodil, his kitchen-woman. Bodil, daughter of a hill clansman, grew simples in the palace garden and collected wild herbs from her native valleys, an hour's walk to the west.

Roderick's daughter, the princess Geruth, found her father's obsession with the old lady tiresome, but then she found little in palace life to excite her interest. Her husband, the nobleman Horvendile, bored her, and his more interesting warrior brother Fengon was always away somewhere fighting. When Geruth bore a son, Horvendile the second, or Hamlet as the nobles named him, after the fashion for new things that swept through the court, Bodil was suddenly summoned by the king (against Geruth's pleading to keep her away) for a special audience. The boy was windy. He wouldn't sleep. His crying kept everyone awake.

'It's his grandfather's troubles, come down in the blood,' said Bodil with her secret smile. 'Give him to me.' And from then on nothing passed young Hamlet's lips that had not been blended and strained in Bodil's kitchen.

The boy grew stronger, though he had fits of melancholy and talked in his sleep.

He gratefully fastened his filial love on his grandfather's kitchen-woman, and rarely visited his parents' wing of the castle.

Then everything changed.

Roderick was killed in battle and Horvendile was crowned king.

Geruth and Horvendile were happy in the first years of their reign, and both having large and uncomplicated appetites, they ordered the herb garden re-planted with table vegetables. Geruth occasionally asked Bodil for wild strawberries and caraway, but these were small exercises for her talents, and Bodil grew irritable from under-use. She spent her time drying out hill herbs while the summer sun was high.

When Fengon returned from his latest foray, he took one look at Hamlet and recommended riding and fencing to put some fire in the boy's blood. Fengon spoke frankly to Geruth about what he saw as failings in his nephew. A future king, Fengon told her, should be putting on muscle not sitting in the kitchen with a cup of milk.

Geruth and Fengon often walked together in the orchard, and beyond the palace walls when Horvendile was busy with court matters.

In Fengon's company Geruth found her sense of humour again. She lost it, she told him, raising the boy. The experience was so taxing she took precautions not to conceive again. In their conversations, Geruth and Fengon often speak about conception, power, how to breed good from good, and what qualities are desirable in a strong sire. It was no surprise to old Bodil to come upon the queen and her brother-in-law lying in warm grass together, out of sight of the palace, trampling, she noticed wryly, a perfectly good patch of cinquefoil.

When it was time to send the boy away to school Bodil had orders written for his food and medicines. She packed his favourite treats into hampers and supervised their loading. When the cavalcade passed through the gate she went to her room to weep. Watching over this were Geruth and Fengon. Geruth now subscribed to all of Fengon's opinions about her son and it was her ruling that Bodil remain behind at the palace during the boy's schooling.

Hamlet was gone only a matter of weeks when a flurry of changes overcame the palace. Geruth began to wear her hair in a new way and asked for honey and goat's milk in place of venison. Fengon gave orders that were once only a king's prerogative, and Horvendile, relieved of many burdens of state, grew fat from too much toasting and loud, long meals that went past midnight in the banquet hall.

It was explained to Bodil by Geruth that she was old and in need of more comforts, so a well-furnished set of rooms had been made ready for her, as befitted a faithful retainer. Bodil's possessions were carried, with minimum courtesy, to the remotest wing of the castle.

Soon a new figure was seen striding the parapets with Fengon and his great hound. From her window Bodil watched Fengon pause and bend his ear to a thin man, whose curled black hair and sun-browned colour marked him as a visitor from the south. The groom confirmed her suspicions. The party, originally from Italy, arrived late one night, with haughty airs and incomprehensible commands regarding the stabling of their horses.

Bodil made it her business, now that she had nothing to do but warm her bones, to watch the southern man. Granted every privilege by queen Geruth, the visitor appeared to covet nothing but privacy. He walked alone in the garden and, when he thought himself unobserved, the ground where the palace waste was thrown. Bodil saw his gloved hand stroke the plants that grew in the stinking pile of rubbish. Over several days he plucked many bell-shaped flowers and hairy leaves, concealing them in large pockets inside his cloak. Bodil questioned the cook who said the Italian was boiling plants over the fire in his room. He ate none of the cook's food, and hadn't been seen in the company of the king, even when the hunting party brought in a fine stag.

The corridors and inner rooms of the palace became places of whispering and intrigue. Stories were carried to Bodil with her evening food. It seemed that everyone, except Horvendile, his face red and bloated from carousing, sensed danger. The birth of a five-legged foal confirmed the ostler's worst fears.

Saying she needed to collect herbs, Bodil carried her basket into the valley to a place where stinging nettles grew. Wearing gauntlets she struck out at the weeds, tearing and stamping them to fit in her basket, then walked home through the forest, stopping to pick mushrooms.

Bodil judged the day of the attack on Horvendile exactly.

Since the full moon, Geruth had been courting her husband with glances and kisses and suggestive words. Horvendile had lost some of his wits to the ale but he had never lost the desire to lie with his beautiful wife, nor given up hope that she would turn to him again one day. Geruth had the orchard swept and a bower made in the fashion said to be new in Italy. She had roses and peonies and mint strewn on a blanket, and invited her husband to pass the long twilight by her side. The best wine was brought from the cellars, and Horvendile's favourite foods set out to tempt his stomach.

Fengon strode into the orchard, coming upon them as if by surprise and complimenting his brother on having such a wife. Horvendile was flattered and offered his brother a place at the banquet but Fengon excused himself with a deep bow to Geruth.

Very late in the evening, Horvendile fell asleep and Geruth stretched him out on his side. When he was snoring loudly she signalled to Fengon and the Italian. Bodil watched the Italian take a vial of black liquid from his cloak and the three conspirators bend over the king's body. A gloved hand poured the poison into Horvendile's ear.

She saw Fengon and the queen slip away together, and the dark man stand a while watching the sleeping king. After some minutes, the Italian turned and walked towards the hedge where Bodil was hidden. He passed without seeing her, his cloak brushing her own.

Bodil tried to rouse the stupefied king. She rolled him onto his back and dribbled the stew of nettles into his mouth, inclining his head to the cup. The king shuddered, gasped, thrashed his arms and fell quiet. Bodil put her ear to his chest. The heartbeat was meandering but strong. He would live.

All night she put the cup to his lips, a sip at a time, dipping into her small wooden bucket with the patience of a mother at a sick child's bed. At dawn when the king was shaking off the last of his terrible nightmares, Bodil roused the guard.

The king has fallen, she reported, and must be put to bed.

Queen Geruth met the news of Horvendile's fall and recovery with a face like a glacier. Fengon publicly ordered prayers of gratitude and a sacrifice, and privately had the Italian and his party killed.

Horvendile took this fainting episode as a warning from the gods to mend his habits. He fasted, gave up ale, and founded an orphanage.

Hamlet returned from school to find his father looking younger than when he went away. Fighting had broken out on the border and Horvendile was relishing the return of his powers. Hamlet noticed his mother's grieved looks, and his uncle's bad temper. Only old Bodil was as he had left her.

When the border fighting turned to war, Horvendile led his men into battle and was killed with honour.

Hamlet assumed the throne in his twentieth year, and married a pale girl from the provinces. When she died in childbirth he quickly married again.

His second wife Else was kin to his mother Geruth's brother, and her character was irritable and demanding. She soon tired of her husband-king and his love of writing poetry. When Geruth's brother plotted an uprising, Else betrayed Hamlet and he died in his thirtieth year from a clean sword thrust to the chest.[48]

❯━┥━◆❯━O━❰◆━┝━❰

Pouring henbane into the ear seems a strange, primitive, awkward business. How could they be sure it wouldn't leak out? How much needs to be absorbed for it to 'work'? Why not spike his food? Was Claudius any more or less paranoid than other kings about being poisoned at his dinner table? Did he go through a lot of poison-tasters? Culpeper says 'the oil of the henbane seed being dropped in the ear is good for deafness, noise and worms.' Killing kings doesn't rate a mention.

❯━┥━◆❯━O━❰◆━┝━❰

The striking snake (like the flying dart and jabbing hypodermic) is the quickest way to get poison to its target. The whole churning apparatus of the stomach is bypassed; the venom soaks straight into the blood where it is taken on a high speed and reckless ride.

To cure the bite of the snake, one needs imagination and timing. And perhaps faith.

In old Malay culture, for instance, snakebite was treated by cutting the whiskers off cats, burning the whiskers (not the cat) to ashes and mixing them with opium. In India peacock tail-feathers and opium were combined in a similar hot mix.

In the frontier days of white American civilisation, a snake-bitten man was given a gallon of whiskey to drink, or had his bitten foot plunged into the uncooked carcass of a chicken to draw out the venom. Tourniquets were always considered useful, as was cutting a good lump of flesh from around the puncture, supposedly to stop the venom spreading

Most cultures where people and poisonous snakes share an environment have favourite remedies which enter the folklore because, as it turns out, only a small percentage of snake bites result in death, and survivors' memories tend to fasten on what was unusual (like washing the bite in toad's urine) at the expense of the real heroes of the day — support, common sense and luck.

With the coming of counter-poisons, in handy ampoules; the analysis of the composition of venoms; the nutting-out of how each one mangles the body's physiology — and how to block that disruption — another layer of comfort was added to our chances of living through envenomation.

Provided we want to be rescued.

These are cases, mainly suicides, where rescue is the *least* desired outcome, where determination *not* to survive the cup of death is our strongest amulet against failure. Cleopatra for instance, who brought off her near-perfect death by eliminating interference; and Sylvia Plath, who laid her head in a coal gas oven on a night when she knew no one would call.

Out-foxing the rescuers, the children of Isis, is part of the dying game.

There's a story about the Egyptian sun god Ra and his granddaughter, Isis, goddess of the Nile. One day Ra was bitten by a snake and despite his divine powers he became deathly sick. He called for Isis, who was known across the land as 'the great sorceress who heals'. Isis mixed seeds, juniper berries and honey to make a balm, but before giving her treatment she demanded to know the secret word which was the source of all Ra's powers. This was a terrible price for the sun king to pay but, as Isis pointed out, he had no choice.

The secret word given to Isis was *hekau*.

Isis poulticed the wound and to make sure her cure worked, spoke the magic word. Ra recovered, and Isis basked in her new status.

But there is more to the story. The serpent who bit Ra had been made by Isis herself, out of soil and some of Ra's saliva. She moulded the wet earth and chanted over it until, as Ra's shadow passed across its outline, it came to life, biting him on the ankle.

Ra made a treaty with all serpents forbidding them to bite ever again, but snakes being what they are, they soon returned to their natural ways. So Ra gave his magic word to a north African tribe of snake charmers, the Psylli, and charged them with keeping serpents at bay.

>—+—»—·—O—«—+—<

Here is my last rescue fantasy, for an embattled queen:

Aziza was the daughter of a snake handler. She had one brown and one green eye, a freak of nature that could be turned to profit. When she turned twelve her father initiated her into the secrets of his trade.

On the day of her initiation, Aziza bathed all perfumes and oils from her skin and hair, and dressed in fresh clothes. Her father called up the spirits by chanting, first in low tones, then worked himself into shrill, unearthly exclamations. Satisfied that the spirits were awake, he took a snake from a sack by the door and spoke to it in whispers. Moving in close, he put the snake to her ear where it bit sharply, bringing a swelling, then a small drip of blood onto her shoulder. He milked blood from the puncture and mixed it with dirt from the floor, spitting to make a paste. He rubbed her ear with the paste, then made her open her mouth so he could blow his breath inside. The ceremony closed with a prayer.

All this attention was flattering to Aziza but in her secret heart she wanted more from life than to be feared by the ignorant classes. Soft cloth and gold suited her tastes more than pennies thrown by foreigners. She wanted a position at court.

So, with bribes and a few threats, her disappointed father made her dream come true.

At the palace she was given menial work until the oddness of her eyes caught the notice of Iras, Cleopatra's lady-in-waiting. Aziza, who knew how to exploit a chance, invented superstitious stories to go with her exotic stare and from the first moment she saw the queen, fell deeply in love with power.

All through the palace there was whispering and panic about the approaching Romans. Cleopatra alternated between calm and despair, one minute threatening to raise an army, the next falling on her divan in tears. Her maids were exhausted trying to keep up with her constant, often contradictory, demands and in the confusion Aziza seized the chance to insinuate herself. She took up a post by Cleopatra's side, never leaving her for a moment, hardly sleeping, alert to the smallest change in her mistress's mood. The girl soon formed the view that these high-ranking women, from the queen down, had no idea about fighting and winning. The senior maids, raised in comfort on good food, would starve to death on the streets where Aziza came from; they were far too ready to fall in with everything the queen said, and totally devoid of the warrior spirit. What angered her most was this talk of suicide. She had never met a defeatist attitude before. She begged the maid Charmian to call in her father. He knew everyone in the city; he knew who could be bribed, who could be bullied, and he had contacts that stretched to Rome. What he could do for a queen was beyond thinking. But Charmian dismissed the girl's ideas out of hand. There was too much to do preparing for the after-life, so just as suddenly as Aziza had risen to a position of favour, she was put back in her place.

The very next morning Iras took Aziza aside and said, 'Today you die with your queen. Make your preparations.' It was an order, not a request, and the girl was made to put on her best clothes. In the afternoon a servant brought them fruit. Aziza knew the boy, his father was a farmer with a mistress in town. He kept his eyes lowered, though she knew he would later brag that he'd seen everything.

Iras lit incense. They prayed to Osiris, then Charmian nodded to Iras. After kissing the queen, Iras threaded her hand into the mass of grape leaves and garnish as if she was looking for something. Aziza strained to see. She already knew there were snakes in the room.

The scene became a parody of Aziza's initiation. While Iras squealed like a schoolgirl, Charmian pulled out a cobra, the drowsiest specimen Aziza had ever seen, and tried to make it bite the queen's breast. The thing lay there like cooked eel. Finally, it bit.

'Now you, my dear,' Charmian said, pulling Aziza out of the way of Iras's flailing arms. Aziza was mortified. She took her bite on the forearm and pretended to faint. Iras was still having a bad time of it, but the other two were willing themselves into the afterlife with prayer cycles. A few seconds of this and the girl could stand no more. There were noises outside, men arguing.

For the first time in her seventeen years Aziza no longer cared about offending the older generation. If they wouldn't listen they could learn. She bent over the queen, who was nicely laid out on her golden throne ready for Osiris, and started pulling at her robes. Finding the marks she began to suck, tasting the blood for poison. Sucking and spitting, she moved from hand to breast to hand. Aziza blew air into the queen's mouth, and with her own bloody lips pursed in imitation of her teacher, began to ululate her father's strange words.

The gods in their mercy heard and obliged. Cleopatra woke up. Aziza helped her to sit, then stand.

This is how they were found when the door burst open and Octavius's men entered, swords drawn.

Bloodied, disarrayed, kohl running down her cheeks, Cleopatra swayed in the arms of a young girl whose face gave off a triumphant, tear-stained light.

Fully awake now, Cleopatra demanded her looking glass. When she saw Aziza's handiwork she slapped the girl full in the face. At the queen's command the guards dragged Aziza outside where she was stabbed to death.

Alone and dishevelled Cleopatra faced Octavius without the benefit of maids.

>―+‹›‐○‐‹›+‹

When illness turns into trouble for the larger society (by being untreated, ignored, denied, vilified, left to spin off into altered behaviours) the social order is disturbed. This is the medical ethnology model which links illness to a sort of aggression or crime against society.

By removing himself from society through suicide, Vincent van Gogh fulfilled his end of the bargain as a 'sick' citizen. When he couldn't be 'fixed', he shot himself.

Vincent wrote to his sister Wil, 'Every day I take the remedy which the incomparable Dickens prescribes against suicide. It consists of a glass of wine, a piece of bread with cheese and a pipe of tobacco. This is not complicated, you will tell me, and you will hardly be able to believe that this is the limit to which melancholy will take me.'

Today we call melancholy by its generic name 'depression', and we know that (at a neurotransmitter level) persistent, organic melancholy is a result of serotonin deficiency.

Vincent's sort of melancholy has been described variously as bipolar episodes, artistic temperament, epilepsy and sunstroke. Today we zoom in on the root cause of melancholia with drugs called 'selective serotonin reuptake inhibitors', the theory being that if the serotonin deficiency is topped up, then brain cells switch back into sunny mode, burning off the clouds . Or we use ECT, or counselling, or institutionalisation, or hope.

Each generation, shaped by different ideas, picks over Vincent's attractive bones, drawn to the tragedy of the tortured artist, like theatre season ticket holders. Everyone knows it will be a good show. One scientist has even tried to rewrite the script.

What if it wasn't depression? What if the painter was slowly poisoned? A university professor from Kansas

City wrote two journal articles in which he argued a link between drinking absinthe (to excess) and the slow chemical poisoning by a component of absinthe called 'thujone'.[49] The professor based his argument on an incident of irrational behaviour by the artist one day close to the end of his life, when Vincent drank turpentine (which comes from pine trees) straight from the bottle. Thujone (from pine tree sap) and wormwood gave absinthe its distinctive, bitter taste (and kick).

Jan Hulsker, Vincent's biographer, argued against this theory.[50] In order to achieve toxic levels van Gogh would have to have been a seriously addicted, chronic absinthe drinker, and Hulsker's primary sources cannot furnish a shred of evidence to support the assertion. Paintings done of Vincent sitting behind a glass of the green alcoholic drink, and paintings Vincent did himself of a carafe and a glass, simply express, writes Hulsker, the 'preferences of the impressionist painters'. Vincent himself said he suffered from epilepsy, and, when he was under examination in the Lunatic Asylum at St Rémy, named members of his family who shared the disease, which may be true or may have been a coded reference to a less well-defined malaise which afflicted all the van Gogh siblings — the mute agonies of depression.

><->-O-<->-<

Antidotes come out of the shadows, like kind spirits holding lamps. The light may be as sharp and precise as a laser; the dying body might suddenly sit up in an excited fever of returned vitality, but, in the main, much of what we can do for lapses in our mental and physical health is ordinary, like van Gogh's wine and cheese.

Chapter Twenty-three

Kraken dreams

I WENT LOOKING FOR poison and found a maze in a garden. All the trees looked benign, the grass lay sweet and green, and the wildlife took no notice. I saw the stark edges of things, the definition and geometry. I found fruits and flowers, garlands, and lots of women cooking. Men sunned themselves. Children tortured ants on the path.

At the centre of the maze, a priestess rested her fingers in a green pool and pondered. I watched jealously, wanting to learn.

Trying to get closer I turned a corner. The path curved and crossed a bridge.

Voices called me. I went towards a scene that looked like a painting. Young men in smoking jackets complained of boredom. In the absence of sofas they reclined against tree trunks, breathing in and out.

An old woman held out a cup and asked me to drink. I meant to decline but I drank, and died.

I wanted to die well, like a heroine. I forgot that the

dying body disgraces itself. My mouth fell open. My eyes rolled back.

The young men lit cigarettes. One of them turned a page of his book.

In the vastness of the garden my dead body had no more meaning than an autumn leaf.

Finding me cold and sufficiently still, the old woman set fire to an owl's feather and blew smoke into my nose.

The gardener gathered my feet and spun me like a caber. Released, I flew through the air and fell into a pool. The priestess welcomed me as a diversion from reflection gazing.

When she got bored she changed me into a toad. My waxy skin and big eyes were useful against hunters. I hopped when I felt restless but mostly I was faithful to the priestess.

One day when the old woman wasn't looking I jumped into her stew.

She died of eating me.

━ ┼ ◆ ━ ○ ━ ◆ ┼ ━

Dreaming about poison is, I suppose, an ego-free expression of my conscious wish for an end, or at least a resolution, to years of wondering. The dream opened out like a banner: silky, rippling, emblematic. If only it had been so easy to find this *place* where the answers were concealed — was it a metaphor like the priestess's pool? or, as I confidently believed when I began hunting, a real country with all its signposts straight and honest.

The rational mind gets tired of reading for meaning. I sent it on a holiday one day and let my senses play over a book of very old maps, envious of the cartographer's

freedom to invent his monsters because there was no one
to say he couldn't.

My poison country, I decided, would resemble the
conjectured Islandia, a place beautifully drawn and
coloured five hundred years ago, before the New World
and its white coats denied the existence of krakens.

On the Islandian west coast with its wind-blown jigsaw
of inlets I could put a clan of sorcerers who harvest the
land and sea and need little more than the elements and a
few seeds saved from good seasons to manage their
secrets. Their concerns would be hermetic, parochial and
fiercely territorial. Plants would figure prominently in
their pharmacopoeia, as would the ritual drinking of bull's
blood during initiation, raids on villages to the south for
brides, and exaggerated reverence for ancestors.

The sunny Islandian south coast and all the hinterland
(comprising nearly fifty per cent of the island's surface
area) would be royal estate lands. At a tradesman's entrance
to a back door of the palace there is a brisk trade in love
philtres and poisoned candles, courtesy of the physician's
assistant who knows just enough about medicine to be
dangerous. Members of the royal household live in fear of
poisoned meals, and there are no less than twenty poison-
tasters at the banquet tables on festive days.

Travelling north-east from the palace you would come
to the small hamlets famous for their witches, sworn
enemies of the sorcerers to the west, and obedient
servants of their forest gods. There are literate women
among them — exiles from palace scandals — and a
lapsed priest or two. Their recipes are complex, and some
ingredients require visits to graveyards and orphanages.
During certain phases of the moon it is wise for villagers
to lock their doors.

In the far east, on a piece of land too poor to farm, would be the only university, a monkish building with laboratories that are, frankly speaking, little better than the fire-fed cauldrons of the witch counties. But the scenery is unobtrusive and the constant wind keeps scholars from stagnating. In one of the lakes a scientist keeps a young kraken specimen caught by fishermen. His plan is to study its habits and keep the king informed.

Watching over the island is the careless deity whose fat cheeks send brutal winds to terrorise the countryside in winter.

The terrain of this old world is harsh. Life expectancy is low. Only travellers with botanical interests will reap any benefit from landing on its shores.

Why choose a land like this one? Why not something closer to home, more modern? The answers go to the heart of reading my way through libraries of words on the character and habits of sojourners who stumble into, or deliberately book a passage to, the territory of poison. The younger me who learned to acknowledge the presence of an unseen menace in empty bottles is the same me who is confronted by the conundrum of physical evidence overlaid by, or informed in some way by, an engagement with a force I can't see. And there is a strong desire in me *to see*, to engage, to comprehend what it was that the natives waited so patiently for while their poisonous brew heated over the fire. When I revisit the romantic image of the cauldron I have to remember that the image is lifeless, inert, until someone speaks. If we go back to that image, or choose another one, say *An open place. Thunder and lightning. Enter three witches,* there is symmetry and purposeful composition to the picture. The cauldron, the fire, the brew, the hand that stirs the pot, the all-important voice that speaks the spell. Perfectly balanced, this structure falls, spills, ceases to have its numinous extra meaning, if the last of these compositional elements is missing. The invocation acts across the other elements like an enzyme that sets the system in motion. Without a voice to sing the spell, how will the spirits know to assemble? If the witches had thrown their fenny snakes and dog's tongues into a pot but been frightened away before singing their chant, the whole thing would have cooled to a rancid broth fit for nothing.

So, my imaginary land has space and tolerance for spells, other ways of knowing, and breeding grounds for strong emotions (succession, power) and an emerging science that hasn't yet got the upper hand.

>-+-◆--O--◆-+-◄

New world science is uncomfortable with terms like 'witch' and 'love philtres'. Romantic language belongs to novelists and dramatists who are licensed by society to publicly explore magic. Science prefers to deal in verifiable facts. A subject like poison lore with its contending aims, questionable methods, vague and often irritable conclusions can be one thing to storytellers and quite another to assayists. And the world has moved on. The great perceived dexterity of the current age is its ability to analyse and interpret data. In its hands, the old cartographer's map has been redesigned to look like a flow chart or an ECG trace or an academic paper with footnotes.

⊱┄❖┄○┄❖┄⊰

Since the middle of the twentieth century, poisoners have moved to the suburbs. Their poisons are different; they get them in different and faster ways. If you want, you can go into a pharmacy and stock up on codeine-based cough medicines; you can download the formula for speed off the internet and cook it up in your kitchen from jumbo packs of sinus tablets. Or go doctor-shopping for Serepax and opioid analgesics which your pharmacy will home deliver.

There are airports in this new world where drugs arrive in the body orifices of 'stuffers' and 'packers' and the witches who make up your brews belong to highly organised gangs. On the street you can buy just about anything. Boiling up oleander twigs, or stewing hemlock in the crockpot, are anachronistic hobbies for eccentrics.

The map of the poisoner is another story. So much depends on the historical period, and who's drawing the picture. In the Indian Shastras written about 600 BC a poisoner 'does not answer questions, or gives evasive answers; he speaks nonsense, rubs the great toe along the ground, and shivers; his face is discoloured; he rubs the roots of the hair with his fingers; and he tries by every means to leave the house.' It's a graphic picture. You get a sniff of caste prejudice in its language.

There are many who would draw the map along gender lines. Women did it. Women got up to tricks under the guise of cooking and nursing and matchmaking. Others would make a case for education (the well-read poisoner), still others for politics.

After years of looking at these categories it seems harder to keep the groups apart, to see what one is asked to see as characteristic. Instead of differences and exactness, one is struck by what is the same in each character, and in each crime. Hundreds of years after Paracelsus's now discredited definition that everything is a poison, only the dose makes the difference, the same might be said of people — everyone is (no, that's too categorical), everyone has the potential to be a poisoner, only the circumstances make a difference.

>--+--<>--O--<>--+--<

Consider this story of a circus elephant that died mysteriously one ordinary night in 1946. Wirth's was a popular, big-top act touring Australian towns with a standard corps of performing animals, including a valuable African elephant insured for £500.

The police, finding no evidence of external injury, arranged for the local doctor to dissect out the animal's stomach (standard procedure when poison is suspected), pack it in a drum and send it to a city pathologist — a not inconsiderable task given that the organ is about the size of an inflatable dinghy. Great care was taken not to contaminate the stomach, or spill any contents.

Suspicion fell on an unknown outsider, a punter with a kink or a grudge, who might — for all anyone knew — try to kill their other valuable exhibits. Given that a large dose of poison would be needed to bring down an elephant, speculation about a killer with a hoard of dangerous toxins spread from the circus tents to the town and beyond. Soon there were reporters mingling with the troupe, and a guard was posted on overnight watch.

When word came back that the pathologist had found the unmistakable signature of arsenic all over the elephant's stomach, the search for the guilty party had a direction in which to turn. The means and opportunity were established (anyone with agility enough to jump a fence, it seemed, could have approached the elephant at its tethering post on the night it died); the final piece of the puzzle, the motive, might take a little longer to fathom. Suggestions that it was an insurance job were never taken seriously.

Between them, the police and the circus boss put together a scenario which made sense of the incomprehensible. A man, under cover of night, carrying a bucket of arsenic and straw, crept past the sleeping troupe in their caravans, dosed the elephant and vanished into the darkness.

A great deal of speculation gathered around this image of stealth in the night and, indeed, the missing space between midnight, when the keeper went to sleep, and 6 a.m. when he found his beast dead, is like a velvet cape that every hand wants to stroke, and some want to wear.

The secret poisoner creeping about in the night with his death brew lives in many imaginations. I imagine the troupe sitting around telling the story again and again, and someone remembering a shape pass by a window; or hearing an unearthly groan, that might have been a lion, but on reflection could have been 'him'.

For the story to hold its suspense, it needs to end here. Humans love mystery. For now, consider the story finished, add it to your favourite unsolved crimes.

⊱┈◆┈○┈◆┈⊰

On my regular drive home from work I cross a bridge then swing right and follow the bank of the Brisbane

Water, a broad, silky sweep of river that is often pepper-mint-green at dusk. A couple of times a year a circus sets up in the park facing the water, and one evening I saw an elephant tied to a pole outside a tent. The poisoned elephant story was fresh then, and my mind was alive to the dangerous outcomes of leaving a beautiful beast (actually, this one looked a little old and shabby) exposed to the random forces of human cruelty. I stopped the car and walked as close as I could, taking in every detail. Before long I began making up alternative stories to the real one: stories which circled back to the time before the circus came to town, stories that attempted to confuse the sequence and reverse the outcome.

Chapter Twenty-four

Out of the coroner's court

JUST WHEN I HAD ALMOST abandoned hope of finding documents that might fill in the gaps in William's story, several pieces of paper miraculously dropped into my hands. I wondered about this. Was my eye sharper, was I more practised, less numbed by hours of sitting in libraries turning the pages of old newspapers, deep enough into my topic to think intelligently about who might report a small incident like the death of a child?

In a regional newspaper from 1927, I found my man: a reporter who covered the inquest into Patrick's death, a young man on the way up, I told myself, a man with a fresh eye and a feel for the narrative. In his opening paragraphs he sets the scene: a small audience of spectators, including neighbours, and 'sisters of the deceased boy's mother'. The room is cold. There is coughing.

My reporter has obviously been chatting to one of those neighbours whose opinion of Macbeth is unshakeably negative and suspicious (or, it may have been Rose), and whose nudge at his elbow made him take particular notice of Macbeth's apparel. He writes:

> Dressed immaculately in a grey morning suit, William Macbeth, father of the deceased boy, a doctor of naturopathy, herbalist and author of a treatise on *Premature Decay* told the court that he had purchased a bottle of Easton's Syrup for his wife who had stated she needed building up after her recent confinement. Easton's Syrup, it was stated, contains iron, quinine and *strychnine*. It is in common use and may be purchased from any chemist. Mrs Macbeth took one dose and left the bottle on a pedestal near the bed. When she returned to the room, she found the bottle missing and at once roused her husband, saying 'Paddy has swallowed the syrup'.
>
> A search was made of the house. The boy first stated that he'd drunk the syrup but later asserted that he had thrown the bottle in the fire.
>
> The child appeared normal and sat by the fire singing. Mr Macbeth did not think anything was amiss, but to make certain decided to ring up Dr Allan. At this juncture, little Paddy stiffened in his chair. His father carried him to a bed, and at once rang Dr Allan, who gave instructions to administer a salt and water emetic and bring the child to the surgery.
>
> On returning from the telephone, Mr Macbeth found that his wife had already taken the child to a neighbour's residence, where the latter was in the process of administering an emetic.

Reading this now, I can see Ellen, a tigress with a threatened cub, over-riding William's 'it'll be all right' attitude. Her imperative is to *do something*. Her instinct is to run to a woman who will act while men make telephone calls and pronounce.

The report continues: 'The distracted father then took the mother home, leaving the boy to the care of the neighbour. He was absent a couple of minutes, and on returning the child appeared to be collapsing.'

I showed this sentence to my sister and asked, 'What do you think is happening here?'

'He was hiding stuff, and giving Ellen a good talking to.'

This sister is six years younger than me. She's familiar with the many versions of Patrick's death and more comfortable than I am with a suggestion that Ellen complied in a cover-up.

'She had to,' my sister insists, 'women *had* to in those days.'

For the record, William made no mention of his patent medicine business. The apparent inconsistency of him purchasing a bottle of medicine when he had dozens of boxes of tonic stored at the house remains a puzzle.

'She probably wouldn't drink anything he made,' my sister comments, 'would you?'

'So you think Patrick really did drink Easton's?' I ask.

My sister shakes her head emphatically. 'No. Not a chance. He drank the home brew and William thought up the Easton's story to cover himself. He put Ellen in the hot seat, and face it, she'd just had a baby so the pieces fitted together — blood loss, iron tonic — and because of her country girl's fear of courts and processes, she was skewered into an impossible situation. On oath, in front of witnesses.'

'Why didn't the coroner tease all this out?' I wondered.

'Someone has to *suggest* foul play.'

We looked up a coroner's responsibilities in a library book. Later in another book I found the crime writer, Dashiell Hammett's, wisdom on the subject (from 1929): 'The current practice in most places in the United States is to make the coroner's inquest an empty formality in which nothing much is brought out except that somebody has died.'[51]

'The coroner,' my reporter concludes, 'had no doubt that the boy died from strychnine poisoning, innocently self-administered.'

>—+—•>—O—<•—+—<

I want to freeze-frame this image of the shattered couple walking out of the coroner's court into the cutting August winds, each party to a deception which may or may not hide a simple, explainable truth, had they the courage to utter it.

My reporter takes one last look at the Macbeths, arm in arm, William with his head high, eyes straight ahead; Ellen's face hidden by a handkerchief and a cloche hat, her slim figure neat and fashionable in a fur-trimmed coat.

>—+—•>—O—<•—+—<

Dorothy Parker called the twenties and thirties the 'smartypants era'. Men of a particular kind, born around the turn of the century — the ones who survived the War, and those who didn't attend — threw themselves into the self-inventing party atmosphere of a time when a bit of talent, a bit of hard work and a lot of front paid dividends. William, I believe, was one of these bucca-neers. He tried to shake off his inherited values and beliefs (and his Catholic education) but they sat so solidly at his core that after the madness of the thirties he aban-doned his disguises and drifted back to his clan. By then he'd caught TB, was thin and wasted by tobacco and drink, short of money and burnt out. Jane, as always, was there to take him in.

But I've fast-forwarded, and I want to see him as he was then, in 1927, when he behaved like someone granted immunity from consequences.

There are no quotes from William, no echoes of his own voice; and try as I might I can find no 'treatise on premature decay', whatever that may be. Logically, I suppose, it fits in with an advertising campaign for his aphrodisiac tonic and may have been printed on flyers which have not survived (if they ever existed).

What William chose to reveal of himself in this public outing before the coroner and how he chose his words are crucial to any understanding of the man and his alleged crimes. My reporter doesn't quote him directly. He paraphrases an account of events which is lucid and believable but which, it seems to me, markedly changes its tone at the point where 'the boy stiffened in his chair'. There are many ways to read this shift in emotional tone. Either William is angered by having his judgement called into question, or worse, he's been genuinely frightened that a doctor and inevitably a policeman would be called to the house; or he's distressed because he's accidentally brought about a disaster, that somehow he's failed to anticipate the tragic consequences of meddling with poison in a house where children play.

Ellen is harder to read. Hidden in her fur collar and
bell-shaped hat, she seems to be running blindly down a
foggy street. There is something terribly wrong here,
something unbalanced in the rhythm of her footsteps.

Patrick was the *second* child she'd lost in a year. Any
woman who suspects her husband of murder *twice* and
sits quietly through a legal hearing — no matter how
numbed she might be by loss — must be mad or guilty or
dangerously biddable to another's wishes. If we unfreeze
the image of Ellen on the court steps and let her hurry
into her future life up to the point four years later when
she takes William to court to fight for property, when
accusations fly from her lips with nothing held back,
when she gives the whole story to a reporter from the
Truth — 'all he earned he spent in drink and I had to get
money from the State' — nowhere does she level the
killer shot, the death blow, her ace that trumps all
bidders: 'My husband poisoned our sons.'

The difficulties I've had understanding the poison
story might be read off that hidden face, if Ellen would
only look up and meet my gaze. The key to apprehending
Ellen lies, I made up my mind, not in Patrick's death but
in her first major grief, the loss of Thomas, the boy who
Rose believed shouldn't have lived past birth.

Chapter Twenty-five

As children do

THE FATE OF THOMAS has stubbornly refused to fit its shape into the narrative gap. For someone like me, peering into the story from the sides, trying to see over the fence, past the looming figure of Rose with her garden hose ready to shoo me away, there's something not quite right. A child's death is universally tragic. Strong emotions are unleashed. 'Thomas died', says the story, and the family did what? Where was the uproar? a diary entry in Jane's book? Nothing. Instead, we hear doors closing, see clothes packed away, appliances and a wheelchair returned. Then silence.

All the digging I'd done into records had produced not a single word connecting Thomas to the house in Katoomba where he is supposed to have died. Several searches for death certificates yielded nothing, due, I now realise, to the change of surname. I'd spent too little time on Thomas, and began to wonder if I'd absorbed some sort of unconscious prejudice towards him. Fleetingly the thought struck me that he might be alive somewhere, locked away in a home, a forgotten old man trapped in a wheelchair.

Had I overlooked the handicapped boy to mourn for and rescue his sweet-faced brother?

Imagining the child Thomas Macbeth has been a reconstructive exercise, completely different from the way I framed his brother and father. The Thomas I conjured inhabited a painting not a wheelchair. This is an odd leap of imagination because the portrait that seemed to sum up Thomas, Edvard Munch's *Puberty*, has all the wrong visual clues. The subject is a seated girl. She's naked and watchful. Her shoulders incline forward in a protective way. To her left is a tear-shaped mass, like a black misshapen shadow or emanation. What I see, and what always disturbs me when I stare at her, is the sense of imprisoned anguish in her posture. She seems to embody the loneliness that existentialists tell us is our inescapable lot; at the same time she projects an awareness, even a defensiveness about her physicality. That awareness is a long stretch for Thomas, who Rose says was like a little animal; nevertheless, I shape the boy this way, as a private allegory.

><•>—◦—<•><

Very late in this story, almost at its end, I picked up my father's birth certificate once again. Date of birth, 23 July 1927. Siblings of the newborn: three living, none deceased.

When I read those words the first time, I put the emphasis on *none deceased*. It created a space in which my father has a role in the story. He is the portable child at the breast, the tiny winter baby carried off by gipsy parents who'd overstayed their welcome. He is also the child who supplants the timid Charles as the baby of the family, and possibly (anything is possible) escalates the displaced Patrick's tantrums.

Reading the sentence in the light of ten years' familiarity with events, I was suddenly struck by something so obvious I was shaken by its impact. *Three living* is the way to read that entry; emphasis on the first two words. Three living means three sons, Thomas, Patrick and Charles. Thomas is *alive* in July 1927, three months after he is supposed to have been poisoned by his father.

Alive but tellingly absent. By July 1927 he has dropped out of the story without trace.

Reinvigorated, I paid for searches and waited. As before, nothing came back and gradually I forgot about the boy, or kept him in a part of my mind waiting to be activated.

My self-appointed task, as I always saw it, was to present possibilities. They went something like this — (i) William Macbeth murdered his sons, or (ii) didn't murder his sons — someone else did it — (iii) the boys killed themselves, accidentally; or (iv) they just died, as children do.

Thomas's vanishing fell between books on a shelf. I couldn't see it any more, so the question didn't exist. Then, almost six months after I'd posted my search, the second miraculous piece of paper fell out of the sky (via the Registrar's Office). 'Thomas Macbeth, son of William and Ellen, died 1934, aged twelve years, of pneumonia while an inmate of a mental institution, on an island on the Hawkesbury River, north of Sydney.'

The effect of this document, with its authority and stark notations, was, initially, a sense of shocked relief. In 1934 William and Ellen had been divorced for three years;

the children were living with Ellen and her new husband; William was spending his way through another man's fortune, high on the high life. William didn't 'kill' his son. He wasn't even there. Half of his crimes were wiped out in a single sentence.

It took a few seconds to feel the aftershock.

Who carried the burden of Thomas's secret disappearance? Whose idea was it to say he was dead?

I haunted the records office, wrote to and rang the asylum, and after a month the file I'd been told was 'lost' turned up. A friendly female voice on the phone cautioned me about the pre-enlightenment terminology I'd find. I shouldn't let words like 'cripple' and 'idiot' worry me or my father.

Tucked into Thomas's dossier are handwritten letters from Ellen to the superintendent of the asylum. Typed on the back of each letter is a copy of the superintendent's reply.

How was I to read this correspondence against the overpowering force of my preconceptions? How do we, the family, deal with this irrefutable evidence of William's innocence, at least where his first son is concerned? Is the story that Thomas died of pneumonia any better or worse than having him die of strychnine?

Death brings closure, and, ultimately, silences the talkers. 'Thomas died' was apparently easier to say than 'we locked him up in an institution'. Such statements open up an uncomfortable dialogue, leave a messy ending that is no real ending at all, just a pause. So, into the place where the central truth of Thomas's fate should sit, a fiction was inserted, because, we can only assume, the missing truth must have been so unpalatable, so shameful, that being murdered by his father —

the counter-narrative — was preferable.

(And yet 'the passionate devotion of mothers to imbecile children', as Leonard Woolf describes it in his memoirs, runs counter to Ellen's actions.[52] Woolf's experience in English villages where mothers kept these children at home, at the expense of other siblings, convinced him that there was an 'unconscious sense of guilt and desire [for the mother] to vindicate herself and the child'. Only with wild protest do these mothers endure their children being put into homes, and, in one case, Woolf accompanied a mother on a rescue mission to get the boy back.)

But I am running ahead of the story. There was a whole file to read.

Thomas Macbeth was given up to the State of New South Wales, under the Lunacy Act of 1898 — Fourth Schedule, in August 1926, and signed in by his mother Ellen. William's name appears nowhere in any documentation. The medical examiner writes his précis: 'No sign of intelligence. Spasm of limbs when he tries to move.' The Wassermann test is negative. He is three feet six inches tall, weighs forty-nine pounds, he does not speak.

There are no letters from William in Thomas's case notes. Thomas, I suspect, ceased to exist for his father the day they took him away. It would be misleading to read anything sinister into this willing forgetfulness, and I don't attempt it. I point instead to the times when mad 'others' were routinely locked away, either in institutions or in attics out of sight. There are examples enough in the archives of lunacy (even the British Royal family tucked away an embarrassing relative in a comfortable 'home'); and literature of course is fascinated with the idea, from Rochester's first wife onwards. I know a man who believed himself an only child, only to discover in later life a brother locked away in a home for so-called 'incurable epileptics'.

Backtracking, I imagine the sequence like this. The household is straining to stay afloat and keep Thomas at home. Ellen falls pregnant with a fourth child. The pressure escalates. The decision is made to do something with the boy, and the options, when they are laid out, are frightening and irrevocable. Nevertheless the boy is assessed, and given up. One day in August Ellen makes the journey with her first-born and leaves him, like a parcel, at the door of the edifice that will feed him and care for his needs, but cannot and will not love him as its own.

The carrying out of this very difficult decision, to commit their son forever (there was no way back from a committal under the Lunacy Act), became for whatever reason Ellen's task. Ellen put herself and the boy on a train to Newcastle in order, I believe, to finally deliver him. Their fraught connection and the terrible elongated attempt to separate him from her body (the umbilical cord choking the oxygen from his brain just a minute too long), the four years of caring for him, the toll on the family, would finally be severed.

After it was over Ellen took up her life again. She wrote letters to her family telling them Tommy had been taken to a special hospital. When everyone was used to the new arrangement, the Macbeths began telling the lie that would reverberate for seventy years; they said that Tommy was dead.

Ellen's first letter to the asylum, written two weeks after his installation, says: 'Dear Sir, Would you kindly let me know how my son Thomas Macbeth is, and I will be greatly obliged to you.' The superintendent replies: 'Thomas has been rather better the last few days. He has been running a temperature. There is however, I am glad to say, no cause for anxiety and his progress is satisfactory.' Just enough information to satisfy a mother. Exactly enough detail for her to connect the words on the page to her intimate knowledge of her son's stricken body; for her mind to work over how and when he might have caught a chill, and what they might be doing to keep him healthy.

A month later the superintendent's reply reads, 'Thomas is making satisfactory progress. Though he is still rather pale, he takes his food well, sleeps well and appears to be contented.' Again the detail, four pieces of information for Ellen to ponder and explore. For the next three years the timid requests for information (and notification of the Macbeths' frequently changing addresses) make their way to Newcastle, and to each letter the superintendent replies in the positive, interspersing his 'satisfactory progress' reports with details of bronchitis, fevers and minor ailments. Then after three years the line goes dead, the letters peter out, the connecting thread loosens its tight grip. (I'm reminded of my own periods of intense involvement in other countries with other families, and how it's impossible to keep it going once you return home. The intimacy seems to survive about six months of writing or phoning then loses its freshness and relevance to the immediacy of 'now'.)

So what happened to Thomas? After seven years at Newcastle Mental Hospital he is three pounds heavier than his admission weight, still can't sit up or speak, requires 'every attention'. The superintendent decides to transfer him to Rabbit Island, a final destination for incurables. He sends a letter to Ellen at her last known address; it comes back 'return to sender, not known here'.

Thomas's notes from Rabbit Island show that he had regular elevated temperatures for the next nine months (climate change? natural decline?). In September 1934, he developed pneumonia. The manager sent an urgent telegram: 'Thomas Macbeth seriously ill. May be visited.' The telegram was returned 'the addressee is not at the address given & is unknown'. Thomas died the same day.

>-+◦-○-◦+-◄

In 1927 Ellen and William first uttered their lie of convenience. Three months later their second son Patrick drank poison and died. The sequence cried out (at least, I believe, in Ellen's mind) for an hubristic interpretation: the counterfeit death was a lightning conductor, the angry gods made a twitching bolt of fire and hurled it, like a javelin, at Patrick.

When Ellen buried her face in her fur collar, she was hiding terror, not tears.

Chapter Twenty-six

Rumour of a midnight monster

THE ENQUIRY INTO the death of the circus elephant didn't end on a note of Gothic speculation about a midnight poisoner. Narratively, cinematically, stories have a duty to descend from the excited peaks of suspense to the lower ground of common, shared experience, and while there are limits to our capacity to understand the motives of the director, there are no limits to the ways we can mentally restack the deck in favour of better outcomes.

We might argue that the elephant should have been on an African plain with its family, not hauled away by thugs and sold to the circus where its wild ways had to be tamed, its instinct to be with its own kind subdued, its dumb wants and needs reduced to a bucket of straw and water. Or if we accept its circus fate, we might give it a kind keeper. By day it does its tricks and gets applause, by night it's fed and bedded down. (Considering that a mature African elephant needs fifty gallons of water and three hundred pounds of vegetable matter a day,

I wonder that it had time to do any tricks.)

The elephant story doesn't, however, end in mystery. We are back in the territory of the schoolteacher and the poisoned well — which is no accident as both stories are taken from the same casebook and involve the same meticulous, unromantic investigator, a man with no time for shadows or spectres, and the determination to turn every stone.[53]

The stomach and its contents are crucial to arsenic analysis after death. Elephants are herbivores with well-known habits. In the wild they spend up to sixteen hours a day foraging, tearing down tree limbs, snatching up grasses, fruits, vines, shrubs; in captivity, they depend on the ingenuity of the keeper, and the abundance of grass near their tethering poles.

During his examination of the stomach contents the pathologist found a surprising omission among the mixture of apples, chaff, straw, salt, water and gravel: not a single blade, not even a smear, of grass. He telephoned to ask why. The answer came back from the keeper that the ground near the hitching post had had no grass on it to begin with: nothing he could do about it, just the way it worked out once they'd set up and pitched camp. You took what space you were given.

Next, the pathologist asked the local police to question the community about strangers who had passed through in the last twelve months — travelling salesmen, tinkers, tourists — anyone who had any connection with, or might have camped on, the ground where the circus set up. The inference the police took, that everyone took, was that they were looking for the poisoner in another disguise.

The pathologist meanwhile was looking at the gravel

in the elephant's stomach. When the police filed their list of likely suspects the pathologist pointed confidently to two travelling salesmen who had come to town a few months before the circus and advertised a remarkable demonstration — the perfect poison for weeds. They had drenched the ground with liquid and invited everyone to see the results in the morning. Overnight the weeds died, the impressed farmers bought a few bottles and the travellers moved on. But what was the missing link? How do we jump from weedkiller to dead elephant? Mischief needs stepping stones.

An elephant, explained the pathologist, scratches the bits it can't reach by sucking up dirt and gravel and blasting it over its stomach and back. It sucked up several yards of arsenic-coated gravel, swallowed some in the process, and died. He could show them a pattern of marks on the stomach lining, if there was any doubt in their minds.

⤛ ⬦ ○ ⬦ ⤜

Is the circus story useful in teaching us who is to blame?

Everyone and no one.

Everything and nothing.

If we had a culprit the ending would be neat. *Crazed elephant-hater apprehended today.* We could all throw stones.

Stories don't always have to make sense, or deliver a moral. But where would we be if every story ended at Act IV Scene I of a Shakespearian play — *A dark cave. In the middle, a boiling cauldron. Enter the three witches.* Despite the neat wrap-up given by the pathologist, the rumour of a midnight monster persisted. We seem to require our myths and monsters, as a way of putting evil in a place we can see it — out there, outside us.

I feel sad for the elephant. Others think it's better off dead than being kept in a circus.

><+>-O-<+><

Grimm has a story called 'The Widow'. She sits all day and night grieving for her two dead sons. Going to church one morning she finds a strange light shining and all her dead relatives sitting in the pews. An old aunt says to the widow, 'Look toward the altar, and you will see your two dead sons.' She looked and saw two children, one hanging on the gallows, the other broken on the wheel. 'See,' continued the aunt, 'thus would it have happened to them, had life been given to them, instead of their being mercifully taken by God when they were innocent children.'

The widow is grateful for God's clemency, returns home, lies down on her bed and dies.

No one in my family is grateful that two boys died in childhood. But the thought has to be considered that Thomas's parents gratefully gave their son — who was living a kind of twilight death, somatically alive, breathing, eating, but on every scale of measurement, far below normal — into the care of the State. They *allowed* the idea that Thomas had died to develop a life of its own.

><+>-O-<+><

I now believe that Rose knew the truth about Thomas, if not from the beginning, then certainly after she opened her doors to her sister in flight from her broken marriage. Sisters thrown together by tragedy and need share a covenant. They tell each other everything, they bare their souls. Ellen, I believe, couldn't have kept her secret, would have inserted the old grief of Thomas into the newer grief for Patrick.

If Ellen had lived, the story of William's sins would have been singular not plural. One death, Patrick's, sheeted home to him. One only. Once Ellen died all bets were off. Rose owned the copyright.

⤐⎯⟐⎯⟐⎯⟐⎯⟐⎯⤐

I visited my father last year carrying one of those spiral-bound folders with A4 plastic inserts. All the documents relating to Thomas, from his admission to Newcastle to his death on Rabbit Island, were set out in order, photo-copied and enlarged for easy reading.

The letters from his mother and the superintendent's replies I put into another section, deeper in, so that someone casually opening the folder wouldn't meet the frontal assault of those raw, grieving petitions.

My parents lived in the country then but were packing up the acreage after fifteen years of self-sustaining hard work for a small cottage near a lake, not far from my house.

The folder ticked like a bomb in my overnight bag.

I suggested a walk and my father and I struck out for one last look at the high view to Oberon. We fell into the easy noticing way we had with each other — one of his ewes needed dagging, there was another tree down — and we both puffed on the steep last bit of the climb. Where we stopped near the fence the full sweep of green pastureland came into view, and with it a bite of mountain wind, cool and definite.

He began a story about breeding ferrets from the days when he lived with Mrs C, the old lady. I don't know where it came from, some connection he'd made from fences to rabbits to ferrets, but there it was, unfolding out of the past.

'They're designer pets now,' I said, not ready yet to relinquish the present, wanting a bit more time to lay the ground for my bombshell; still, down the lanes of the past we went as I learnt the trick to getting a cranky ferret's teeth off your finger. When my father pulls out a snapshot from his past, it's usually wholesome and instructive, and nearly always gently funny.

'I thought the old lady bred greyhounds,' I said.

'Greyhounds, ferrets, turkeys, ducks, anything that would turn a shilling.' He swung around to look across his fifty acres, then back to the view. 'We milked morning and night, and there was the vegetable garden . . .'

As he listed the frantic productivity of Mrs C's small farm I suddenly understood what I'd never understood: why they'd moved from a comfortable city house to the pioneer lifestyle of a working farm. I saw what he'd been trying to recreate, re-animate, be the master of. And how he'd made his move too late — as strong as he was (still is), he's older than Mrs C was back then, and more tellingly, he's had no unpaid labourers to do the back-breaking work.

Breaking into his recollections, I told him, from start to finish, about the file on Thomas, the whole thing in one continuous sentence, ending lamely 'so you see . . .'

It might have been my imagination, but I thought the

cosy bond between us slackened a little at his end, though he was still with me and listening.

'We'd better see what your mother's up to,' he said finally, frowning into the wind.

⊱—⊹—◦—⊹—⊰

Over the washing up I asked my mother if she thought Dad was up to reading letters written by Ellen.

My mother lost two brothers when she (and they) were young; one to leukemia, one to a car accident. But these were out-in-the-open bereavements, processed in the normal family way of solidarity in loss. She said she'd talk to him.

And so, with much preamble, I handed over my piece of solid research, with accompanying photographs of Rabbit Island, then and now. There was television noise from another room, and an unnatural absence of noise around our table. My father, I realised, was not as eager to accept my gift as I was to give it. It was like sitting in a darkened theatre with a curtain that wouldn't go up.

Forcing the point, I opened at the first page and walked them through it. 'See, this is the report from the first doctor at Newcastle Mental Hospital, see the date —'

And so on. I gave up under the strain after two pages.

My father is neither dull nor discourteous. He has been, all his life, a generous and genuinely interested listener; life and people delight him. Yet, confronted with my folder of facts and jottings and documents, he stumbled. The tension between his position of disavowal and mine of avid testimony (the familiar battleground of historical witnessing) did not yield any common ground over the dinner table. We stared at our hands, the vase of flowers — anything, it seemed, but my folder.

Later that evening, in the pre-bed, cocoa-drinking phase of the evening, I asked him if he *at least* wanted to read his mother's letters. I couldn't let it go.

I took his silence as a yes, a defeated sort of yes, and efficiently flipped through the pages finding the place. I switched on a reading lamp and left him with the book open on his lap. I said goodnight to my mother who was already in bed, then doubled back to the sitting room, expecting to find the scene I'd composed in my mind, my father in tears re-living the lost years. Peering around the corner I saw him with the remote control, flipping through television channels with the sound off, my folder carefully closed and put to one side on the coffee table.

>‑+‑＜○＞‑+‑＜

Thinking over that evening now, writing the story, I'm worried by my apparent incivility. My presumption that my version of events had more validity than anyone else's, that feelings should be made subservient to the urgency of new information; it felt like the juvenile can't-wait interruptions of a teenager whose grasp of the immediate is so strong it pushes out the past and future.

Some 'facts', I've come to see, are too potent on their own, too stark. If you lay them out and leave them sitting there — like the death card in a fortune-telling deck — unqualified, bald, out of any context, you inflict something awful and unexamined on the unsuspecting other.

My father *did* read his mother's letters (and possibly communed with his own invited ghosts), when it suited him. He assimilated the true fate of his eldest brother into his private memoryscape. Together, after he'd moved, we speculated on the sort of hardships Ellen must have endured trying to keep Thomas at home as long as she did. But we kept William out of the picture. We were back in our comfortable denying relationship, full-circle to William's sample case.

〉—◦—〈

Missing sons have to be accounted for. In Katoomba Cemetery one day in 1999 I found Patrick. There's no marker left, only a space in a sunny plot with a north-facing aspect. Apart from my sister, who consulted the Council records for me and tracked down the cemetery map, I think I'm the only person who's spent time at the graveside since 1927. I can't think of Patrick as a potential uncle. He never grows past three years old, and is always a pretty boy sitting in front of a fire.

Thomas is buried in a common grave at Brooklyn Cemetery, on the Hawkesbury River. It's impossible to say exactly where since the bulldozers and landscape architects beautified the site. But it pleases me, in a perverse sort of way, that Thomas, the awkward piece of baggage, unclaimed even in death, lies under lemon-scented gums in expensive real estate with breath-taking water views.

〉—◦—〈

One could make a case for multiple crimes against these two boys. Revisiting the interview with Rose in 1980 in the light of this new information, I am struck by the old lady's audacity. She assumed that I would take her testimony at face value, possibly believing that the past couldn't be disinterred and not crediting me with the will to do the digging. Rose (who, despite everything, makes me feel protective) sized me up at that interview as yet another woman with advantages denied to her and played me like a fish on a line.

This woman deliberately lied to me. The difference between saying that one boy was sent to an asylum, the other swallowed poison, and saying their father poisoned them, is monumental: it's not just untruthful, it's vindictive, and unaccountably sad.

Ellen and William don't get off scot-free either. They were an accident waiting to happen. The demons are said to congregate around certain couplings. I look at this pair — fire and water, city and country, domestic and wild — and see the omens. They weren't Scott and Zelda, but their journey tracked in the same time frame, with a similar hedonism and a spark that lit up briefly then burnt to ashes. William's cousin (I learnt much later, from following up Jane's genealogical tables and notes) was at Princeton with F. Scott, and was the source of some of William's more outrageous claims about his education and American connections.

Herb gatherers in ancient times took magical precautions when they picked poisonous plants. Magic, if you believe in it, can aspire to the divine (the so-called 'high magic' of Orpheus), or to the immediate world (the pantheistic or low magic which connects everything to everything else — people to planets, herbs, stones, metal and so on). To protect themselves from the deadly mandrake, the plant whose tap-root looks like a stalking demon, the Greeks drew three circles in the earth and cut the plant (which was said to scream) while facing the west. This is the secret of all magic practices, the thing that makes it contrary to the laws of God: magic is an attempt to exert power through process; do this and good will happen, do that and beware.

><+<>+<>+O+<>+<>+<

I'm still thinking about what happened to Thomas. Every time I drive south I glance across at Rabbit Island. Once I drove in across the land bridge for a closer look, and quickly left again, stared down by what I read as the understandable contempt of the inmates.

I think about what happened to Patrick differently on different days.

The day I drive to the house in Wilson Street where the tragedy happened I feel nothing. I watch a man and a woman each holding the hand of the child between them walking up the steep incline at the top of Wilson Street, towards the bus-stop on the main road. They're not well dressed in the way my mother would call 'neat and tidy', they're more post-hippy verging on feral and they have wide open friendly faces. I spy on them from my car. I judge them to be about the same age as William and Ellen were seventy-odd years ago.

Two things occur to me. The ordinariness of walking to a bus-stop; and the separateness of myself as voyeur, watching and making up stories about what this or that means. Take his dreadlocks and nose ring. Or the fact that it's Tuesday and he's not at work.

I notice that the woman, who started out in a playful mood, is now tight around the mouth and eyes. Something has upset her. The man and the child take no notice of her change in mood, they continue with their counting game. You could read a lot into that. If, tomorrow, I heard that the woman had killed the man, or the man had smothered the child with a pillow then shot himself, I would have my opinions already half worked out.

But no such thing happens, and I'm left to watch them waiting for a bus.

On other days I feel a sort of existential grief for all the people and elephants and cats and rats that have ever died a tormented death by poison, and then I know I'm getting a touch of the romantics. When I particularise that grief to my father's brother, I want to go back to Wilson Street on an ordinary day in 1927, knock on Dr Macbeth's door, and, like a good daughter of Isis, see what I can do to keep the day ordinary.

THE END

Endnotes

1. HIATUS PHRASE See Peter O'Connor, writing about the watershed moment in a story, in *Dreams and the Search for Meaning* (Sydney: Methuen Australia Pty Ltd, 1986), pp. 95–96.

2. CIRCE Circe the enchantress lived on the island of Aeaea. When Ulysses and his crew landed on her shores she used her magic to transform the crew into swine. Ulysses, armed with an antidote to Circe's poisoned cup, overpowered her magic, then succumbed to her charms and lingered at her banquet for a year.

3. MANY VERSIONS OF THE LAST HOURS OF CLEOPATRA For a brilliant analysis of the life, death and myths of Cleopatra see Lucy Hughes-Hallett, *Cleopatra: Histories, Dreams and Distortions*, (London: Pimlico, 1990). For a more conventional version, see Carlo Maria Franzero, *The Life and Times of Cleopatra* (London: Heron Books, 1968).

4. JOHN GLAISTER, *The Power of Poison* (London: Christopher Johnson, 1954).

5. C.G. JUNG, *The Archetypes and The Collective Unconscious*, CW9, para 158.

6. LIKE PSYCHE HOLDING THE LAMP Psyche was a mortal princess

whose beauty offended Venus so much she sent her son Eros/Cupid to slay her with a poisoned arrow. Instead (as H.A. Guerber retells it), Eros, stung by one of his own arrows, fell in love with Psyche and devoted himself to protecting her from Venus. They declared their love under cover of night, Eros making Psyche promise that she wouldn't try to see his face. *If thou once shouldst see my face/ I must forsake thee: the high gods/ link Love with Faith, and he withdraws himself/ from the full gaze of knowledge.* Psyche agrees, but is tortured with curiosity, and, goaded by her jealous sisters, takes a lamp to see Eros's face (she also carries a dagger to stab him if he's a hideous monster, as her sisters predict). When she sees a handsome youth lying there she's so surprised and happy she accidentally tilts the lamp so that a drop of burning oil falls on Eros, stinging him awake. Enraged, he flies away, saying (as so many have said) *There is no love without faith.* For her sins, Psyche suffers various indignities but is ultimately saved again by Eros who whisks her up to Olympus as his chosen bride.

7. GLENN INFIELD, *Eva and Adolf* (London: New English Library, 1976), p. 242.

8. DIANE WOOD MIDDLEBROOK, *Anne Sexton: A Biography* (London: Virago, 1992), p. 216.

9. DOROTHY L. SAYERS, *Unnatural Death* (London: New English Library, © 1927, 1972 edition), a Lord Peter Wimsey mystery.

10. GEORGE ORWELL, *Decline of English Murder and other Essays* (Harmondsworth: Penguin, 1978).

11. EMMA JUNG AND MARIE-LOUISE VON FRANZ, *The Grail Legend*, trans. Andrea Dykes © 1970 (Boston: Sigo Press, 1986), p. 37.

12. MAHAN VIR TULLI, *Gems Therapy: Precious Gems, Numbers and Colours* (New Delhi: Sagar Publications, 1994).

13. DR BLYTH is my favourite toxicologist. Alexander Wynter Blyth, *Poisons: Their Effects and Detection* (London: Charles Griffin and Company, 1884).

14. FILSON YOUNG, 'Trial of Hawley Harvey Crippen',

Notable British Trials (Edinburgh and London: William Hodge & Co. Ltd, 1919).

15. HANS PETER DUERR, *Dreamtime: Concerning the Boundary Between Wilderness and Civilization*, trans. Felicitas Goodman (Oxford: Basil Blackwell, 1987).

16. T.R. FORBES, *Surgeons at The Bailey: English Forensic Medicine to 1878* (New Haven: Yale University Press, 1985).

17. NICHOLAS CULPEPER, *Culpeper's Herbal or the English Physician*, 1652.

18. JOHN GERARD, *A Catalogue of Plants (1596–99)*.

19. JUVENAL, *The Sixteen Satires*, trans. Peter Green (Harmondsworth: Penguin, 1967, 1974), Satire VI, p. 151. Pontia, daughter of Petronius, poisoned her own children then committed suicide.

20. WILLIAM TURNER, *A New Herball (in 3 parts)*, 1551–68.

21. MICHEL FOUCAULT, *The Birth of the Clinic: An Archeology of Medical Perception*, trans. A.M. Sheridan © 1963 (London: Routledge, 1991), p. 153.

22. SUE WHYBRO, 'The Green Dream', *Australian Police Journal*, Vol. 52, No. 3, September 1998, pp. 170–174.

23. THE BROTHERS GRIMM, 'Death's Messengers', from *The Complete Illustrated Works*.

24. 'DOCTOR DEATH' — this feature article by Susan Chenery appeared in the *Australian* Magazine, Feb. 27–28, 1999, pp. 26–31.

25. ROBERT HUGHES, 'Pre-Raphaelites' from *Nothing if Not Critical* (London: Harvill, 1990), pp. 114–117.

26. REGINA SORIA *et al.*, *Perceptions and Evocations: The Art of Elihu Vedder* (Washington: Smithsonian Institution Press, 1979).

27. JOHN KEATS, 'Isabella; Or, The Pot of Basil', XIII, 103–104.

28. JOHN KEATS, 'Ode on Melancholy'.

29. THOMAS G. GEOGHEGAN: 'An Account of a case of Poisoning by Monkshood, which formed the subject of a Criminal Trial', *Dublin Journal of Medical Science*, Vol. 19, 1841.

30. FILSON YOUNG, 'Trial of Hawley Harvey Crippen',

Notable British Trials (Edinburgh and London: William Hodge & Co. Ltd, 1919).

31. ERIC WATSON, 'Trial of William Palmer': *Notable British Trials* (Edinburgh and London: William Hodge & Co. Ltd, 1923).

32. VOLTAIRE, from *Candide,* trans. Tobias Smollet (New York: Oxford University Press, 1979), p. 120.

33. Though opium-based sedatives are no longer sold, the habit of sedating fractious children with antihistamine — and sometimes codeine-containing mixtures — is often seen in community practice.

34. AGATHA CHRISTIE: AN AUTOBIOGRAPHY (Glasgow: William Collins & Sons: third edition, 1980).

35. PRIMO LEVI, *The Periodic Table,* trans. Raymond Rosenthal (Suffolk: Sphere Books, 1988).

36. STEN FORSHUFVUD *et al., Assassination at St. Helena: the poisoning of Napoleon Bonaparte,* ed. H.T. Mitchell (Vancouver: Mitchell Press, 1978).

37. 'JOHN ERPENSTEIN' from *Murder Did Pay: 19ᵗʰ Century New Jersey Murders,* introduced by John T. Cunningham; with a bibliography by Donald A. Sinclair (Newark: New Jersey Historical Society, 1982).

38. ALAN DOWER, *Crime Chemist: The Life Story of Charles Anthony Taylor, Scientist for the Crown* (London: John Long, 1965).

39. GUSTAVE FLAUBERT, *Flaubert in Egypt: A Sensibility on Tour,* translated and edited by Francis Steegmuller (New York: Penguin, 1972).

40. LOMBARD *et al.* 'Arsenic Intoxication in a Cocaine Abuser', the *New England Journal of Medicine,* 1989, Vol. 320, No. 13, p. 869.

41. OTTO POLLACK, *The Criminality of Women* (Philadelphia: University of Pennsylvania Press, 1950).

42. G. K. CHESTERTON, *The Vampire in the Village,* a Father Brown Mystery.

43. CORAMAE RICHEY MANN, *When Women Kill* (Albany: State University of New York Press, 1996).

44. 'THE WIRREENUN WOMAN AND HER WIRREENUN SON'

in *Wise Women of the Dreamtime: Aboriginal Tales of the Ancestral Powers*, collected by K. Langloh Parker, ed. with commentary by Johanna Lambert (Vermont: Inner Traditions International, 1993), pp. 96–103.

45. ANTHONY HOLDEN, *The St. Albans Poisoner: the life and crimes of Graham Young* (London: Hodder and Stoughton, 1974).

46. SIR WILLIAM HENRY WILLCOX, 'Acute Arsenical Poisoning', *British Medical Journal*, July 22, 1922, pp. 118–124.

47. THE REAL DR JOHN D. GIMLETTE spent twenty years in Malaya. The first edition of his book, in 1915, was used extensively at poison trials in Malaya and was the first ever reference source of potent vegetable substances, new to European medicine. John D. Gimlette, *Malay Poisons and Charm Cures, second edition* (London: J. and A. Churchill, 1923).

48. THE STORY OF HORVENDILE AND GERUTH is creatively adapted from *The Hystorie of Hamblet*, c. 1608, and first appeared in the *Historica Danica* of Saxo Grammaticus, c. 1200.

49. W.A. ARNOLD, 'Vincent and the Thujone Connection', *J.A.M.A.*, Vol. 260, No. 20, 1988, pp. 3042–3044.

50. JAN HULSKER, *Vincent and Theo Van Gogh: A Dual Biography*, ed. James Miller (Ann Arbor: Fuller Publications, 1990).

51. DIANE JOHNSON, *The Life of Dashiell Hammett* (Reading: Picador, 1985), p. 89.

52. LEONARD WOOLF, *The Journey Not The Arrival Matters: An Autobiography of the years 1939 to 1969* (London: The Hogarth Press, 1973), pp. 49–52.

53. THE CIRCUS ELEPHANT comes from Alan Dower, *Crime Chemist: The Life Story of Charles Anthony Taylor, Scientist for the Crown* (London: John Long, 1965).

Acknowledgements

I owe thanks to general and medical libraries in Sydney, Melbourne, London, Cambridge (UK), New York and Boston, all of which will scarcely know I was there in the 1990s, reading my way through the literature of poisoning.

When help was needed with obscure material, I was able to count on Jack Eckert, Reference Librarian of the Francis A. Countway Library of Medicine at Harvard University, and Philomena Connolly, Archivist at the National Archives, Dublin, and I am grateful for their kind attentions.

I was born too late to meet William Macbeth but I know his son well. My father gave me the gift of William's story, and in the rhythms of his storytelling I found my own narrative voice. Hopefully what I have written will give meaning to some of the strange silent passages of my father's past.

To my mother I simply say thank you for teaching me that all forms of creativity are precious and valid.

Friends old and new have listened to me telling poison stories over the years, and their encouragement for and contributions to the writing of this book have been invaluable: I can't thank you all by name, but special tributes go to Margaret Berg, Colleen Henry, Debra Hely, Bill Tibben, Greg Lill, Victoria Ramsay, Injy Tawa, Janet Reinhardt, Loubna Haikal, Jeri Kroll, Trudi Canavan and Terry Taylor.

Susan Hampton, friend and editor, took the threads of my story and taught me her secrets of skilful weaving. I thank her for years of wise and gentle counsel.

At Varuna Writers' Centre, Peter Bishop told me to be a friendly, compulsive presence in my story and I hope I have lived up to his expectations.

For help with art and art theory I turned to my friends, Daryl Ganter, Cynthia Mitchell, Saskia Mitchell, Frank Neilson and Rose Andrews.

To my sisters, Kerry Egan and Machele Jelsma, who were enthusiastic listeners, and my brothers Roy and Graham, this is your book too.

Just when I needed her, I was lucky enough to secure Selwa Anthony as my literary agent. Thank you for everything, Selwa.

To my publisher Nikki Christer and editor Judith Lukin-Amundsen at Picador, my deep gratitude for embracing the book.

And finally, holding all of this (and me) together has been my husband Andrew, the first person to see the shape of my story.